THE QUANTUM DOT

THE
QUANTUM DOT

A Journey into the Future of Microelectronics

RICHARD TURTON

Department of Physics
University of Newcastle upon Tyne

New York
OXFORD UNIVERSITY PRESS

Oxford University Press

Oxford   New York
Athens   Auckland   Bangkok   Bombay
Calcutta   Cape Town   Dar es Salaam   Delhi
Florence   Hong Kong   Istanbul   Karachi
Kuala Lumpur   Madras   Madrid   Melbourne
Mexico City   Nairobi   Paris   Singapore
Taipei   Tokyo   Toronto

and associated companies in
Berlin   Ibadan

Library of Congress Cataloging-in-Publication Data
Turton, Richard (Richard John)
The quantum dot : a journey into the future of microelectronics /
Richard Turton.
p. cm. Includes bibliographical references and index.
ISBN 0-19-510959-7 (paperback)
ISBN 0-19-521157-X (hardcover)
1. Microelectronics—Technological innovations.
2. Semiconductors. 3. Quantum theory. I. Title.
TK7874.T883   1995
537.6'226—dc20   95-3649

1 3 5 7 9 8 6 4 2
Printed in the United States of America

For Denise, Cathlin, Samantha and Benjamin

CONTENTS

PREFACE

THE impact of microelectronics is apparent all around us. Fax machines, compact disc players and children's toys are just a small selection of the applications, not to mention the computer. This book is about the tiny electronic devices which have made all of this possible, and in particular with the basic physical processes which govern the operation of these devices. The branch of physics concerned is called solid state physics, the essence of which is to explain how the properties of a solid depend on the structure of the material at an atomic level. For this reason we begin our journey at the level of atoms and electrons. This leads us to consider the special properties of semiconductors, such as silicon, and how these properties are utilized in devices such as the semiconductor laser and the transistor. Following a short excursion into how transistors are used to form the basic elements of a computer, we look at the process of integrating a large number of these devices on to a single small slice of silicon—a silicon chip. However, our brief is not merely to examine the current state of affairs. As the subtitle of this book suggests, the main emphasis is to consider the direction of future developments in microelectronics. In research institutes around the world scientists are striving to produce new devices which will be superior to the silicon transistor. One of the main efforts is directed towards a continued decrease in the size of the individual devices. The ultimate conclusion of such a scheme is a quantum dot—a tiny box of matter only a few atoms across. Such structures could form the basic elements of 21st-century microelectronics. The title of this book, *The Quantum Dot*, therefore represents the final destination on our journey into the future of microelectronics. Or rather I should say that it represents one of the possible final destinations since there are several other routes that we could take. I will examine some of these other possibilities in the later chapters.

I have had two audiences in mind when writing this book. Firstly, I have directed it at first-year undergraduates in electrical engineering and physics. However, it is not a textbook as such and it contains no mathematics. Instead the principal emphasis of this book is the presentation of a conceptual picture of the physical processes involved. To achieve this I have made frequent use of analogies with situations which are likely to be more familiar to the reader. Secondly, I have tried to ensure that the book can be understood by general readers, even those with little or no previous knowledge of physics or electronics. For them, I hope this book will provide an entertaining as well as an informative read.

The book is intended to be read from start to finish. The main reason for this is that many of the ideas which are introduced at length in one chapter are revisited at a later stage, often in a rather different context.

Any specialized area of interest has its own associated terminology, and microelectronics has more than its fair share. I have endeavoured to minimize the number of technical terms used in the text, whilst at the same time introducing a sufficient number so that anyone who is so motivated can proceed to more technical works. Each term is explained as it is introduced, and a glossary of these terms is also included for reference.

The occurrence of very large (and small) numbers is common in many areas of physics, and solid state physics is no exception. I have written all such numbers in English (using one billion in the now more commonly accepted meaning of one thousand million), rather than using scientific notation. In many cases I have also used a comparison to try to convey the scale of these quantities.

It gives me great pleasure to acknowledge my wife, Denise, and my editor, Michael Rodgers. Without their support and encouragement I doubt whether this book would have actually got into print. In addition, I would like to thank my colleagues, particularly Milan Jaros and Jerry Hagon, for many illuminating discussions and practical assistance over the years.

THE QUANTUM DOT

The Rise and Rise of the Silicon Chip

LEGEND tells of a chess-playing king called Purushottama who was so confident of his ability at the game that he challenged all comers. Finally he was defeated by the wise Pandit. On being asked to name his reward, the Pandit asked only to be given one grain of rice for the first square of the chess board, two for the second, four for the third, and so on. Foolishly the king agreed to this seemingly trivial request. Anyone who has ever tried to calculate the numbers corresponding to the other squares will appreciate how the repeated doubling rapidly produces huge numbers: the number of grains of rice for the sixty-fourth square is a nineteen-digit number, far in excess of the total number of grains of rice in the world.

The phenomenal rise of the integrated circuit, better known as the silicon chip, is due to a similar exponential increase. The origins of the integrated circuit can be traced back to 1959. Developments in this year by Jack Kilby at Texas Instruments and Robert Noyce at Fairchild Semiconductor Corporation laid the foundations for what was to follow. Two years later a circuit comprising four transistors was successfully fabricated on a single piece of silicon, and over the next few years increasingly complex circuits were produced. In 1964 Gordon Moore, also at Fairchild, examined this trend and predicted that the number of transistors on a chip would continue to double every year. This now famous prediction has come to be known as Moore's law.

Just like the Pandit's request, it is hard to grasp the implications of Moore's law. One way to experiment with this for yourself is to take a sheet of A4 paper, rule a line down the middle of it, and then draw a simple design (a letter of the alphabet would suffice) on one half. Now repeat the operation by dividing the clean half of the paper into two and reproducing your design in one of these quadrants. How many times can you repeat the process? Twelve times? Maybe sixteen times if you have a very fine-tipped pen and good eyesight? Anything beyond this would seem to be impossible. And yet Moore's law continues to hold true after thirty years. Granted, the rate of increase has slowed down somewhat so that the current rate of increase is closer to a factor of two every two years. Nevertheless, it is now possible to accommodate around 64 million transistors on a single chip. In terms of the chess board analogy we are at about the twenty-sixth square, with little sign of the progress slowing. How much longer can it continue?

For the semiconductor industry this is the sixty-four-thousand-dollar question (though sixty-four billion dollars would actually be a more appropriate figure!). In the past the increase in the number of devices on a chip has been almost totally due to constant improvements in the fabrication methods, enabling ever smaller devices and more complex circuits to the constructed. Quite amazingly, amidst such a rapidly changing environment, the transistor itself has changed very little. Progress has been made simply by scaling down the dimensions of the individual devices and adjusting certain related properties accordingly. However, several factors (which we will examine in Chapter 5) suggest that this approach may not be viable for much longer. If the trend is to continue into the twenty-first century, a replacement for the conventional semiconductor transistor must be found.

Before we pursue this any further we should perhaps ask whether there is really any need to squeeze more than 64 million transistors on to a single chip. Let us answer this by briefly examining the benefits of integration. The prime incentive for this dramatic miniaturization has been an economic one. It costs little more to manufacture a silicon chip containing one million components than it does to produce a single discrete device. In addition, the chip already contains the necessary interconnections between the components, whereas discrete devices have to be individually connected together to form a circuit, a process which is both expensive and unreliable. This means that an integrated circuit manufacturer who wants to remain competitive has been forced to follow, if not lead the way, in producing ever more complex circuits on a single chip. The other key factor is performance. Small transistors operate faster than larger ones. This consideration is important not only in state-of-the-art supercomputers where speed is of the essence, but also in more mundane systems such as personal computers where tasks such as the display of high-quality moving graphics make huge demands on the processing capability.

These two factors have contributed greatly to the boom in sales in the semiconductor market. This is particularly true with regard to computers. Most computers are replaced not because they are worn out or have ceased to function, but because newer models, giving much better performance, and often costing less than the original machine, have become available. It follows that the semiconductor industry will make every effort to sustain the continually escalating levels of integration for as long as possible. How will this be achieved?

Researchers are currently exploring several different avenues. One aspect which is common to all of these approaches is that in order to understand the physical processes involved we need to have an appreciation of quantum theory.

The quantum world is a strange and deceiving place. Many of the predictions of quantum theory appear to be contrary to our intuitive perceptions. This is because the world that we experience is generally immune to the minuscule fluctuations which occur on the quantum scale. It is only when we consider extreme physical conditions that these rules become important. A typical case occurs when we drastically reduce the dimensions of an object.

A vivid example of such an effect has been demonstrated within the past decade. It is now possible to create tiny crystals of the material cadmium selenide which contain less than a thousand atoms. Each one measures a few millionths of a millimetre across. (In comparison, the diameter of a dust particle is typically measured in thousandths of a millimetre.) The peculiar feature of these crystals is that, although they all have precisely the same composition, they exhibit quite different properties. In particular, the larger crystals are found to be red in colour, smaller ones orange, and the very tiniest (containing barely a hundred atoms) are yellow. We arrive at the seemingly ludicrous conclusion that the colour of an object depends upon its size. Strange as it may seem, this is indeed the case in the quantum world.

These tiny specks of matter are called quantum dots. On their own they seem to be of little practical use, but if integrated on to a chip their unique electrical properties could be harnessed to perform a function similar to a conventional transistor, whilst requiring only a tiny fraction of the space. Creating a regular array of these dots would allow a computer processor many times more powerful than any current supercomputer to be constructed on a single chip.

This concept is unlikely to become reality for many years. The technology for creating these structures is still in its infancy, and radical changes in the layout and operation of integrated circuits will be required to utilize fully the capability of the quantum dot. However, other new devices are now commonplace. In a quantum dot all three dimensions are reduced to the quantum scale, but similar, although less dramatic, effects can be produced if only one of the dimensions is decreased to this length scale. For example, a quantum effect device can be achieved by constructing one or more layers of material which are only a few tens of atoms thick.

Another way to access the quantum world is to produce (through the use of exceptionally low temperatures) an extraordinary state of matter called a superconductor. Recent advances, particularly the discovery of so-called high-temperature superconductors, indicate that in the future the technological importance of these materials may rival that of semiconductors. Among their many potential applications they can be used to make extremely fast electronic devices. A further possibility is to construct devices which respond to beams of light rather than currents of electrons. Such a change-over has already taken place in the field of data transmission where the superiority of optical fibres over copper cables is well established. Now optical devices which perform similar

functions to a transistor are under investigation in research laboratories around
the world. Will these take over from the electronic transistor? Or will one of the
other new technologies prove to be dominant? We do not yet know the answer
to this question, but I hope this book will provide some insight into the
fascinating range of possibilities that lie in store.

1

Nature's Construction Set
Assembling the Building Blocks of Matter

O UR journey begins with the atom, the fundamental building block of matter. It is a fairly safe assumption that the reader of this book is familiar with the concept of a submicroscopic world populated by atoms. The paper in this book, the chair you are sitting on, in fact everything you can see is composed of atoms. We are well aware of the existence of these atoms, and yet this knowledge seems to have little relevance to our understanding of the world on our own much larger scale. Suppose we take a bar of copper. We can compress it between heavy rollers until it becomes a thin sheet, or draw it out so that it forms a fine wire. Whatever we do to it, the bar always appears to behave like a continuous substance. We might conclude that although we know the copper bar consists of small discrete particles, these particles are so incredibly tiny that for all practical purposes we can forget about their existence. If so, then we have a lot to learn from the ancient Greeks.

The idea that matter is composed of small fundamental particles is not new. About two and a half thousand years ago, Greek philosophers were attempting to understand the nature of matter by asking questions such as the following. Suppose we take a bar of copper and cut it into two pieces. We then take one of these pieces and cut that in half, and so on. Is there, in principle, any limit to the number of times we can divide the bar in this way? The Epicurean school of thought, founded on the ideas of Leucippus and Democritus, argued that there was a limit, that at some stage we would end up with a particle which could not be further subdivided. They called these particles 'atomos', or indivisible ones. Of course they had no way of experimentally proving their theory, nor did they try. After all, they were philosophers, not scientists. Nevertheless, they were not satisfied merely to postulate the existence of these atoms. They went on to use this atomic view of matter to account for certain properties of materials. Sour-tasting substances, for example, were believed to contain atoms with jagged edges, whilst small, round atoms were to be found in oily substances.

How does our atomic theory compare with that of the ancient Greeks?

Modern science has left us in no doubt about the existence of atoms. In fact, very recently it has become possible actually to 'see' individual atoms. Using a very fine probe, honed to a point so sharp that the tip itself is only one or two atoms across, it is possible, with the aid of a computer, to generate a map of the surface of a crystal in which individual atoms can be distinguished. Along the way we have discovered that atoms do not really deserve the name 'indivisible ones'. They in turn have an internal structure of their own. However, the Greek philosophers were certainly right about one thing. The idea that the characteristics of solids are governed by details on the atomic scale turns out to be completely correct. Although a copper bar appears to behave as though it is a continuous substance when it is compressed or stretched, an understanding of many of its physical properties—for example, why it is a good conductor of heat and electricity, or even why it is the colour it is—requires us to consider the bar as a collection of copper atoms. It is for this reason that our investigation into the physics of microelectronic devices begins with a look at the atoms themselves.

What is an atom composed of? There are three basic constituents, the electron, the proton and the neutron. The proton and the neutron have a similar mass, the proton having a positive electrical charge and the neutron being uncharged. The electron is tiny in comparison, being approximately two thousand times lighter than each of the other particles. If we use a golf ball to represent a proton, then an electron is just a tiny bead a few millimetres in diameter. Given this disparity in size, it is rather surprising to find that both particles carry the same quantity of electrical charge, the charges being of opposite sign.

The rules for combining these particles in order to make atoms are essentially quite straightforward. Atoms are electrically neutral, and so there must be equal numbers of protons and electrons in any given atom. The simplest such case is the hydrogen atom which consists of just one proton and one electron. Since neutrons have no electrical charge, their numbers are not so strictly controlled. In general, we find that the numbers of protons and neutrons are approximately equal, but the number of neutrons may vary, even between atoms of the same type. For example, carbon has six electrons and protons, and in its most common form also has six neutrons, but an alternative structure exists with eight neutrons. In this latter form, known as carbon 14, or ^{14}C, the atom is unstable and decays by radioactive emission.

Having examined the constituent particles, we now need to consider how they are arranged within the atom. A key step in revealing the structure of the atom was the discovery of the central nucleus by Ernest Rutherford in 1911. The protons and neutrons pack tightly together to form a dense nucleus which accounts for virtually all of the mass of the atom. This nucleus is tiny, even on an atomic scale, being typically a hundred thousand times smaller than the diameter of an electron orbit. This is strongly suggestive of a model of the atom resembling our own solar system. The nucleus takes on the role of the Sun, forming the hub around which the electrons (planets) rotate. It is hardly surprising that Rutherford noticed this similarity and attempted to describe the atom in this way.

Although this picture of the atom as a miniature solar system is very appealing, it was soon realized that such a model suffers from a fatal flaw. It had been known for many years that if a charged particle, such as an electron, is accelerated, it emits radiation. This causes the particle to slow down. In fact the name for this phenomenon, *bremsstrahlung*, is German for 'braking radiation'. Why is this relevant to an electron in orbit around a nucleus? The answer is that even if the electron is travelling at constant speed, the very fact that it is moving in a circle means that it is accelerating. (We experience a similar acceleration, for example, when spinning on a fair-ground ride, or when taking a corner at speed in a car.) The consequence of an electron in an atomic orbit losing speed would be disastrous. It can be thought of in terms of a satellite orbiting the Earth. The satellite must move at a particular speed in order to stay in a fixed orbit, since it is only its motion which prevents it from falling back down to Earth. If the satellite loses speed for some reason, it moves into a correspondingly smaller orbit. If it continues to lose speed it will spiral inwards and finally crash onto the Earth's surface. A similar fate would await an orbiting electron subject to the effects of *bremsstrahlung*. Within a short time, in fact less than a billionth of a second, the electron would fall into the nucleus. Clearly something is very wrong with this model.

To obtain an alternative view of the atom we will take a rather more abstract approach. Rather than worrying about the positions of the electrons it is much more helpful to classify them in terms of the amount of energy that they have. To illustrate this idea let us forget about the atom for a moment and concentrate instead on something more tangible.

Imagine a mine shaft descending deep into the ground. At various depths down the mine shaft there are short horizontal passages leading off to nearby coal faces. It is helpful to have some way of referring to these workings, so we will start at the bottom and call the lowest one Level I, the next one up Level II, and so on. We can define the energy of a particular miner in the mine as being the energy required to lift him from his current level up to the surface. If we assume that all the miners weigh the same amount, then the energy required is always the same for a particular level and increases for levels deeper below the surface. Finally, we will arbitrarily fix the energy scale so that a miner at the surface corresponds to a position of zero energy. In this way if it takes ten units of energy to raise a miner from Level I to the surface, then we say that a miner on this level has minus ten units of energy. Similarly, Levels II and III may correspond to, say, minus seven and minus five units of energy, respectively. We should not be worried about having a negative amount of energy—in fact it

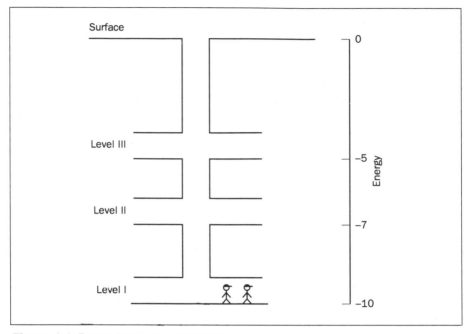

Figure 1.1 Energy levels in a coal mine.

is rather appropriate since it corresponds to a negative height. All it means is that the miner has ten units of energy less at Level I than he has at the surface. This may seem a strange way to do things. Why not start at the bottom of the mine shaft and call this zero energy? Then we would have positive amounts of energy as we move upwards. The reason is that if we have another nearby mine shaft that is slightly deeper, then we may find that the surface corresponds to ten units of energy in one case and twelve units in the other. It is much more sensible to use the surface as a common reference level and measure all the other levels relative to this, even if it means that the energies are negative quantities. If we now draw a cross-sectional view of the mine, as in Figure 1.1, we can also interpret it as a graph where the vertical scale represents energy.

How does this relate to the atom? Each of the electrons in the atom can be assigned a particular value of energy, just like the miners in the coal mine. In principle, we could determine this value by measuring the amount of energy required to extract each electron from the atom. If we say that an electron has zero energy when it is at a great distance from an atom, then the electrons that are attached to an atom have a negative amount of energy. This is referred to as the binding energy of the electron, since it is the amount of energy that an electron saves by being bound to a nucleus. If we measure the energy of the electrons in this way we find that the ones in the outer orbits need relatively small amounts of energy to free them from the atom, while those in the inner orbits require the most. This is exactly what we would expect since the electron is attracted to the nucleus by electrical forces, the magnitude of which decreases rapidly as the separation of the charged particles increases. Consequently, those electrons in orbits closest to the nucleus are the most strongly bound. By following this procedure we can arrange the electrons on an energy picture just like the miners, with the lowest levels corresponding to electrons in orbit closest to the nucleus, and the higher levels to those in successively larger orbits.

This energy picture of the atom may not be as simple to visualize, but it highlights an important feature which is not apparent if we think of the atom as a miniature solar system. We could, of course, construct an energy picture for the solar system in a fashion similar to that above. For simplicity, and to obtain a better analogy with an atom, we will assume that all the planets have the same mass. Suppose that we have an immensely powerful spacecraft capable of dragging the planets out of their orbits and away from the Sun. Pluto, being the most distant planet, would be easiest to remove from the solar system, so it would occupy a high energy level. The amount of energy required would increase as we move inwards with Neptune, Uranus, Saturn, Jupiter, Mars, Earth and Venus requiring increasingly more energy, and Mercury ending up on the lowest energy level. Now let us put all the planets back where they started and make a less drastic change. For example, we might try moving Venus to a slightly larger orbit. Using our rocket, we could push Venus outwards from the Sun by one metre, or a thousand kilometres, or any distance that took our fancy. In doing so we would change the energy of Venus by a corresponding amount. Similarly, we could make adjustments to the orbits of any of the other planets, each time producing changes in the energy level picture.

Let us contrast this with the example of the coal mine. As illustrated in Figure 1.1, two miners start off down on the lowest level, each with minus ten units of energy. We can move one or both of them up to Level II where they have minus seven units of energy, but we cannot move them to a position with minus eight or minus nine units of energy since this corresponds to solid rock. This is quite different from the solar system where we could vary the energies of the planets by any amount we desired. Instead, the miners are restricted to certain discrete values of energy corresponding to the levels in the coal mine. No other in-between energies are allowed.

It was the Danish scientist Niels Bohr who, in 1913, first suggested that a similar description accounts for the electron energies in an atom. Bohr had been working with Rutherford at Manchester University to improve Rutherford's model of the atom. In his revised model, Bohr proposed that an electron in an atom is only allowed to have certain discrete values of energy. What is more, every atom of a given type has exactly the same set of allowed energies. How does this avoid the problem of the electrons spiralling in towards the nucleus?

In Bohr's model of the atom it is not possible for an electron to slow down and move into a slightly smaller orbit since a corresponding energy level does not exist. Bohr therefore postulated that electrons in these 'special' allowed orbits did not emit radiation and consequently were not slowed down. This was not altogether satisfactory since it was not apparent why electrons in orbits around atoms should behave differently from free electrons. To find a better explanation Bohr had to wait for over ten years for the full development of quantum theory, and we shall have to wait until Chapter 7. For the time being, however, we shall make do with Bohr's description of the atom which is more than adequate for our present purposes.

Let us take a moment to examine one important consequence of Bohr's model. Although an electron in an atom is forbidden to lose arbitrarily small amounts of energy, it can move into a different orbit if it gains or loses just the right amount of energy to enable it to make a sudden leap to another allowed level. If it jumps from a high energy level to a lower one it has to get rid of the excess energy by some means. This is done most conveniently by converting the energy into a brief pulse of light. The wavelength, or in other words the colour, of the light is determined by the amount of energy involved. Since only certain energies are allowed for a particular type of atom, this means that each element produces light with a characteristic spectrum. This is something we can observe quite easily—in fact, we witness this phenomenon every night when we see sodium street lights. The yellow light produced is a result of the electrons in many sodium atoms undergoing precisely the same change in energy.

There is one further issue that we need to clarify before we can complete the picture. How do we know which of the allowed electron energy levels are actually occupied by electrons? The position most favoured by an electron is the one with the least energy. Consequently, just as water always flows to the lowest point, so electrons tend to find the levels with the lowest possible energy. Indeed, the reason why our world is populated by atoms rather than separate electrons and nuclei is because the electron has less energy when it is attached to an atom than when it is on its own. (Remember that electrons in an atom have a negative energy compared to free electrons.) We might then expect that all the electrons would fall into the lowest energy level in the atom. However, they don't. This is hardly surprising if we consider the analogy with water. Suppose we pour some water into a glass. The first few drops cover the bottom of the glass, but as more water is added the glass begins to fill up. Similarly, as we add electrons to an atom they are forced into increasingly higher energy levels as the lower ones become full. How many electrons does it take to fill an energy level? The answer is surprisingly simple: just two electrons are allowed in any given level. This result was first noticed by Wolfgang Pauli. It is known as the exclusion principle, for the simple reason that if two electrons already occupy a particular level, then any further electrons are excluded from entering it.

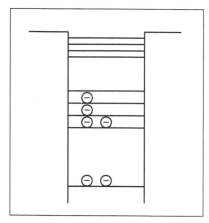

Figure 1.2 A schematic representation of electron energies in a carbon atom (not to scale).

The exclusion principle has far-reaching consequences and we will meet it again shortly. For the time being we will simply use it to illustrate how the electrons are arranged in a given atom. Let us take carbon as an example. We have already mentioned that a carbon atom contains six electrons, and so the exclusion principle tells us that at least three levels must contain electrons. In order that the electrons minimize their energy we would expect these to be the three lowest levels. In fact, for a subtle reason which we do not need to discuss further, it is found that there are two electrons in the two lowest levels, whilst the third and fourth levels are occupied by one electron each. This arrangement is illustrated schematically in Figure 1.2.

So far we have been talking about individual atoms. In nature atoms rarely occur as single entities: they usually attach themselves to other atoms. They do this for the same reason that electrons and nuclei group together to form atoms— because the energy of the group of particles is less than that of the individual particles. Let us begin by considering the hydrogen atom.

The exclusion principle tells us the maximum number of electrons that can be accommodated in each of the energy levels. However, to some extent this can also be thought of as a minimum amount. It is rather like saying that the maximum number of legs that a human can have is two, but on the whole this is also considered to be a minimum requirement. The hydrogen atom with only one electron cannot fill even the lowest energy level. How then do all these one-legged hydrogen atoms manage? Is there some way in which they can get hold of another electron? It is not as simple as just gaining another electron since then the system would be unbalanced, having two negative charges but only a single positive charge in its nucleus. Let us instead consider a pair of hydrogen atoms and examine what happens if by chance they come into close proximity with one another. We can picture each hydrogen atom as a central nucleus with

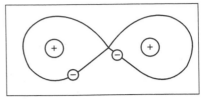

Figure 1.3 A hydrogen molecule. The
electron orbit encompasses both nuclei.

an electron in orbit around it. If the two atoms are close enough, the orbits of
the electrons may overlap. When this happens the electrons can quite easily swap
between one atom and the other. We can now think of this as a single system
where the electrons take up a new orbit encompassing both of the nuclei, as in
Figure 1.3. Although the positively charged nuclei mutually repel one another,
the electrons act as mediators. They can be thought of as the glue holding the
system together. Each nucleus is attracted towards the electrons, causing the two
atoms to bond together. In this way we obtain a system with two electrons, as
desired. Both of the hydrogen atoms are satisfied since each feels as though it
has two electrons. This new system is called a hydrogen molecule. It is a very
stable structure. We can see this since if the nuclei are pushed too close together,
their positive charges repel one another and push the nuclei apart again. On the
other hand if they move further apart, the electrons pull the system back together.
So the actual separation of the two nuclei is very well defined and is the same
in every hydrogen molecule.

The existence of molecules composed of two identical atoms puzzled scientists
in the early nineteenth century. Various chemical reactions indicated that the
basic constituents of many gases were pairs of atoms. This raised some
interesting, and at the time unanswerable, questions. For example, why are there
always two hydrogen atoms in a molecule and never three? Indeed, if two similar
atoms are drawn together then what is to stop all of the atoms in the gas from
forming a single large group? We now know that the answer lies in Pauli's
exclusion principle. Hydrogen atoms pair up in order that the electron levels can
be filled in the most economical way, meaning that the energy of the system is
minimized. Adding a third hydrogen atom would mean that this electron would
have to go into a higher level because the lowest level already has its full
complement of two electrons. Consequently, the addition of an extra atom
increases the energy of the system, and so a molecule of three hydrogen atoms
does not form.

We should be careful not to form the opinion that all molecules consist of
just two atoms. As a slightly more complex example we can consider the methane
molecule which consists of a single carbon atom and four hydrogen atoms. The
lowest energy level in carbon is fully occupied, containing two electrons. Above
this are four very closely spaced levels, as can be seen from Figure 1.2. These
levels contain the other four electrons belonging to the carbon atom, but they
have a total capacity of eight. If we think back to the picture of electrons

executing circular orbits around the nucleus, then the two electrons in the lowest level can be thought of as orbiting around a spherical shell close to the nucleus, whilst the other four are in a second, larger shell. In order to achieve maximum stability the carbon atom must fill this second shell, which means that it needs to gain another four electrons. It can do this by linking up with four hydrogen atoms. By sharing one of the four electrons in the second shell with each of the hydrogen atoms, the carbon atom ends up with a share of four more electrons, and each hydrogen atom with a share of two. In this way all the involved parties are satisfied. Just as in the hydrogen molecule, we can see that the arrangement consists of a specific number of atoms. There is no way that another hydrogen atom can join the cluster because this electron would have to go into a higher energy level. Similarly, if we take one of the hydrogen atoms away, the carbon atom will rapidly acquire another one to replace it.

We began this section by saying that a molecule forms because the energy of the system is less than the sum of the energies of the individual constituents. By examining the above examples we can see that as a general rule this means that each atom does its best to obtain a full shell of electrons. This is because partially filled shells represent spaces where electrons could be placed which would lower the energy of the system, and so it is beneficial to fill all such spaces. In most cases this is achieved by atoms sharing electrons, as described above. We then say that the atoms are joined together by a covalent bond.

*

Although we have only touched upon the surface of the vast variety exhibited by molecules, we must move on quickly as our main concern is with crystals. Molecules may consist of just a couple of atoms or several thousand atoms, but in a crystal the bonds between the atoms link together the whole of the structure, from front to back, side to side and top to bottom. The number of atoms in a silicon chip, for example, is of the order of a thousand billion billion—this could be written as a one followed by twenty-one noughts. How do atoms join together in these sorts of quantities? Remarkably, in many cases they do so by exactly the same process as in molecules, by sharing electrons between atoms.

Before we consider how a crystal is formed, let us briefly return to the methane molecule. A methane molecule has a characteristic shape. What does it look like? Well, stand up, stretch out your arms and legs as though you are in the middle of a star jump and then rotate your body through ninety degrees. Your body now represents the carbon atom and your feet and hands the hydrogen atoms. Alternatively, if you are not feeling very energetic then you can simply look at Figure 1.4. Suppose that we now replace each of the hydrogen atoms with another carbon atom. By borrowing one electron from each of these carbon atoms, the central carbon atom will have a share of four extra electrons as before.

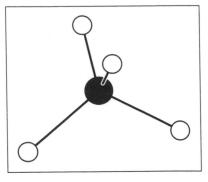

Figure 1.4 A methane molecule. The black sphere represents a carbon atom, while the white spheres are hydrogen atoms.

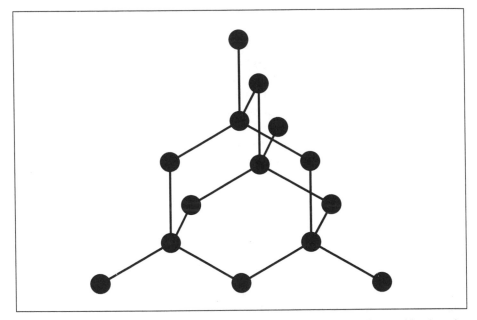

Figure 1.5 The structure of a diamond crystal. Notice how it is built up of basic units which have the same form as the methane molecule (see Figure 1.4).

However, the other four carbon atoms have only five electrons in their outer shell—the four they started with and one that they have borrowed from the central atom. To make up this deficiency they each attract another three carbon atoms. These five central carbon atoms are all now satisfied, but the others still require extra electrons, so they in turn attract further carbon atoms. The process continues ad infinitum so that a large crystal forms. It is just like using a child's construction toy in which each of the building bricks is identical, and in the shape of the distorted star we described above. Each of the bricks joins on to its neighbour in a particular way, as shown in Figure 1.5, to produce a characteristic

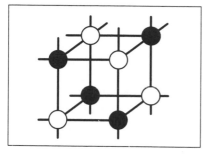

Figure 1.6 A crystal of sodium chloride. Sodium ions are represented by black spheres and chlorine ions by white ones.

pattern which is repeated over and over again. It is this precise regularity across millions of atomic layers which produces the beautiful crystal that we call a diamond. Another crystal which forms in a very similar way, and with which we shall be principally concerned in the next few chapters, is silicon. The similarity is hardly surprising as a silicon and a carbon atom have much in common. With a total of fourteen electrons, silicon has a full complement of two and eight electrons, respectively, in the first two shells. This leaves four electrons in the outer shell, with room for a further four. Consequently, silicon forms a crystal in the same way as carbon, with each silicon atom being joined to four neighbours.

There are several other mechanisms which enable atoms to join together to form a crystal. One example is found in sodium chloride, better known as table salt. Sodium chloride consists of equal numbers of sodium and chlorine atoms. If we look at how the electrons are configured in these atoms, we find that in each case there are two complete shells, containing a total of ten electrons. Sodium has just one further electron in the outer shell, while chlorine has seven, and is therefore only one short of a third complete shell. The simplest way for them both to obtain a filled shell of electrons is for each sodium atom to donate one electron to a chlorine atom. This leaves the sodium atom with only ten electrons, but it still has a total of eleven protons in its nucleus, so there is an imbalance of charge. We have to be careful about the terminology here because the term 'atom' strictly applies only to an electrically neutral body. When the system is no longer neutral we refer to it as an ion, so in this case we have a positively charged sodium ion. Similarly, chlorine has seventeen protons, but there are now eighteen electrons, so this is a negatively charged ion. Since each negative ion attracts nearby positive ions, and each positive ion in turn attracts neighbouring negative ions, we end up with a regular arrangement of alternate negative and positive ions, as shown in Figure 1.6. This process is termed ionic bonding.

Although we have considered covalent and ionic bonding as two quite distinct mechanisms, there is a sense in which they can be considered as two extremes of a single phenomenon. In each case redistribution of the electrons causes a strong bond to form between the participating atoms. In the covalent bond the electrons are shared equally, whereas in the ionic bond one of the atoms gains complete control of the other's electron. In many cases the atoms are held together by a mixture of these two processes. We will consider an example of this in Chapter 6 when we examine the semiconductor gallium arsenide.

The above methods of bonding work very well if the total number of electrons in the outer shells of the atoms add up to the magic number eight, but what happens if we have a collection of atoms each of which possesses only a single electron in the outer shell? Such a situation exists, for example, in the metal sodium. It is not advantageous for the sodium atoms to form a covalent bond since at most each atom can bond with only one other similar atom, giving a total of only two electrons in a shell which could accommodate eight. An ionic bond could perhaps form if seven of the atoms donated one electron to an eighth atom. However, as all the atoms are initially identical it would be very strange if some of them behaved in a different way from others. Instead we find that the bonding mechanism in these materials is quite different.

Since the outer electron in each atom is only very weakly attracted to the central nucleus, it can easily be set free. This means that instead of picturing a collection of atoms we should really be thinking of a group of particles consisting of electrons and ions. Although the ions are all positively charged, and therefore repel one another, the attractive forces between the electrons and the ions are sufficient to hold the system together. The ions occupy positions in a regular array, just like the crystals we have encountered already, but since the electrons are not bound to any particular ion, they are free to move about through the crystal. We will see in the next chapter that this has important consequences with regard to the electrical properties of the material. However, we should note that the electrons cannot escape from the solid altogether since that would leave behind an unbalanced positive charge. It is found that most atoms with one, two or three electrons in their outer shell bond in a similar way. This constitutes the group of materials we refer to as metals.

There is one other principal type of solid, but in this case the basic constituents are molecules rather than atoms. The idea that molecules could attract one another seems quite strange as we have seen that the atoms in a molecule have already achieved a stable configuration. However, although the numbers of positive and negative charges in a molecule are equal, the way in which these charges are distributed causes the molecule to exert a weak force on any neighbouring particles. We can demonstrate that such a force exists by considering how atoms of the inert gas helium interact. Helium does not readily react with other elements, the reason being that it has just two electrons and so has no partially filled shells. These electrons are in constant random motion, and so at some moment we may find that both of the electrons are on the same side of the nucleus, as depicted in Figure 1.7. Under these circumstances the helium

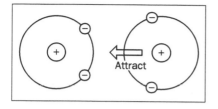

Figure 1.7 At any instant in time the distribution of the electrons in a helium atom may be such that the atom exerts an attractive force on a nearby nucleus.

atom exerts an attractive force on a nearby positively charged particle, for example another helium nucleus. Although the situation is subject to constant change due to the motion of the electrons, the overall force between the atoms is attractive. In helium this force is extremely weak— so much so that helium does not solidify at atmospheric pressure no matter how low the temperature. However, similar arguments can be used to show that there are attractive forces between neutral molecules which are strong enough to allow the formation of solids. In addition, the shape of some molecules is such that there is a constant displacement of the positive and negative charges. It is these effects which cause water molecules to bond together to form ice, and polymers to produce plastics.

We have seen then that nature exhibits a wide variety of methods by which one atom can bond with another. However, one feature which is common to all of these instances is that the central core of the atom, consisting of the nucleus and any filled shells of electrons, plays very little part in any of these bonding processes. In fact, the core is almost completely unaffected by the environment in which the atom is placed. For example, the core of a carbon atom remains essentially the same whether we are dealing with an isolated carbon atom, or one which is part of a methane molecule or a diamond crystal. It is only the electrons in the partially filled outer shell of the atom which interact significantly with the electrons in other atoms. These electrons are referred to as the valence electrons. It is the behaviour of these valence electrons that determines virtually all of the important properties of a solid. In fact, many of the characteristics of a material can be determined simply by examining the type of bonding present. For instance, the sharing or exchange of electrons that takes place in covalent and ionic bonding results in extremely strong bonds. This makes it very difficult to break the bonds either by physical means or by thermal vibrations. Consequently, these materials tend to be very hard and have high melting points. In most metals the bonding is significantly weaker, while the forces responsible for molecular bonding are weaker still. As a result these materials are generally softer and have lower melting points, properties which we readily associate, for example, with plastics. In the next chapter we will examine how the electrical properties of materials are affected by these criteria.

2

To Conduct or Not to Conduct
and
Where Semiconductors Fit In

THE vast majority of microelectronic devices are fabricated from silicon. In the very early years germanium was sometimes used, while in more recent times there has been an increase in the use of gallium arsenide and various other compounds for specialized purposes. The property that all of these materials have in common is that they are semiconductors. So what exactly is a semiconductor?

Let us begin by determining whether or not a given material conducts electricity. We can easily carry out such an experiment with a battery and a light bulb, using a sample of the material as part of the circuit, as in Figure 2.1. If the material conducts electricity the bulb will light, whereas if it does not conduct the bulb fails to light. On the whole we find that metals are conductors of electricity, whereas virtually all non-metals are not.

Now that we know which materials conduct and which do not, we need to ask why it is that the materials behave in the way they do. Let us return again

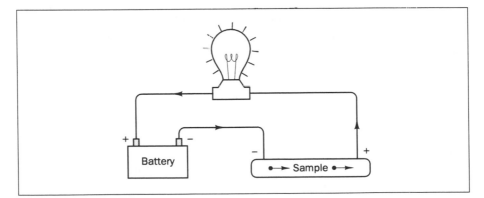

Figure 2.1 A circuit to determine whether a given sample of material conducts electricity. The arrows indicate the direction of flow of the electrons.

to Figure 2.1. If the positive terminal of the battery is connected to one end of the sample and the negative terminal to the other then there is a voltage difference between the two ends of the sample. As a result, any positively charged particles are attracted towards the negative end and the negative charges are attracted to the positive end. The cores of the atoms, consisting of the nuclei and the filled shells of electrons, cannot move in response to this force because they occupy fixed positions in the crystal. This means that the only particles which can take part in electrical conduction are the valence electrons. As we have seen in the previous chapter, the valence electrons in metals are more or less detached from their parent atoms and so are free to move through the crystal and take part in electrical conduction. However, with other types of bonding the situation is very different.

In covalent or ionic materials all of the valence electrons are used to produce filled electron shells, and so none is free to move. A similar result is obtained when the solid is composed of molecules, since all the electrons are required to form the bonds within the molecule.

Consequently, we can conclude that the electrical properties of a material are strongly dependent on the type of bonding that exists in the crystal, and this suggests a clear distinction between materials which conduct electricity and those which do not. Where do semiconductors fit into this scheme? The problem is they don't! The puzzle becomes more intriguing if we remember that the bonding between atoms in a silicon crystal is very similar to that between carbon atoms in a diamond, and yet the two materials have very different electrical properties. Diamond is an excellent insulator, whereas silicon is actually quite a good conductor of electricity at high temperatures. Clearly, if we wish to understand the differences between these materials we need to develop a more sophisticated model. Let us consider how we can do this.

In the previous chapter we discovered that the number of electrons in each state of an atom is strictly limited. According to Pauli's exclusion principle only two electrons are allowed in each of the energy levels. However, we run into problems when trying to apply these ideas to solids. For example, we have seen that a silicon atom has four valence electrons. According to the exclusion principle there is room for eight electrons in this shell as we have four separate states. If this atom bonds with four other silicon atoms then it can fill up these unoccupied spaces, but the system of five atoms now has a total of twenty valence electrons. Where do the rest fit in? Are they forced into higher energy levels? If so, it is hard to see how this can be a more attractive proposition than having five separate atoms. Furthermore, the problem would appear to get worse as more atoms are added to the cluster. Indeed, even for a tiny microscopic crystal we have to think about accommodating many billions of electrons. Are we to assume then that the exclusion principle applies only to atoms? Or can we salvage something of this to make sense of the arrangement of electrons in solids?

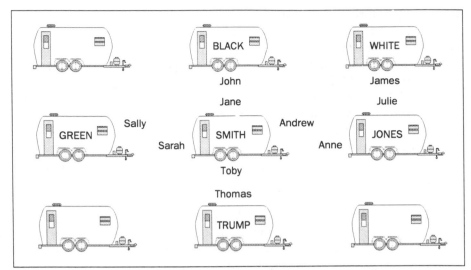

Figure 2.2 The arrangement of families on the fictitious caravan site.

Let us try to answer this by considering an analogous situation. Consider a highly organized caravan site where the caravans are set out in a regimental square grid. The Smith family occupy one of the caravans close to the middle of the site. They are a large family with four children: Andrew, Jane, Sarah and Toby. On the right of the Smiths' caravan is one occupied by the Jones family who also have four children. Their eldest daughter, Anne, is Andrew Smith's girlfriend. Meanwhile, Jane Smith is madly in love with John Black in the caravan behind, Sarah's friend is Sally Green in the caravan on the left, and young Toby gets along famously with Thomas Trump in the caravan in front. Is that clear? If you are still struggling to sort out the happy families then go back and re-read the last few lines in conjunction with Figure 2.2. As you may have guessed by now, the Blacks, Greens, Trumps and all the other families on the campsite each have four offspring. In every case they make friends with children from each of the other caravans adjoining their own, so that the whole campsite is linked together. Now, one day a benevolent millionaire decides to purchase a walkie-talkie for every child on the campsite. He reasons that because there are four children in each family it is sufficient to tune the walkie-talkies to receive and transmit on one of four different frequencies. They are then given out so that each caravan has one walkie-talkie tuned to each frequency. Let us see how they could be distributed, starting with the Smiths' caravan. Naturally, Andrew Smith and Anne Jones are given sets tuned to the same frequency. Similarly, Sarah, Jane and Toby are given ones corresponding to their friends in the other caravans. No problem so far because each pair of friends has a set tuned to a different frequency. Next we can consider the Jones' caravan. Anne's sister, Julie, has a walkie-talkie to talk to her friend James White in one of the other neighbouring caravans. (You may wish to refer back to Figure 2.2 again if you are getting lost.) Obviously, Julie is given a different frequency from Anne, but as we are limited to four frequencies, it must be the same as that used by one of the Smith children. So, for example, Julie will be talking to James on the same frequency that Jane Smith uses to communicate with John Black. This is a rather unsatisfactory arrangement because there would be a considerable amount of cross-talk between the systems. We can resolve this quite easily by re-tuning one of the pairs to a slightly different frequency, but of course the problem does not end there. One of the Green children and one of the Trumps, and, in fact, a child from every family on the site, will be trying to operate walkie-talkies at the same frequency. The best solution is to re-tune them all so that each pair of walkie-talkies on the site broadcasts at a slightly different frequency.

What has this got to do with the crystal of silicon? In the silicon atom the valence electrons have four discrete energy levels, or 'states', from which to choose. When the atoms form a crystal the valence electrons from the different atoms interact and, as a result, change their energies slightly. Since all the atoms in the crystal are linked together, we should get as many different energy levels as there are pairs of valence electrons. Due to the huge number of atoms in a crystal, the differences between the energies of adjacent levels are so small that it makes more sense to think of a continuous band of allowed energies. Accordingly, we refer to this collection of states as the valence band. However, in one respect we need to bear in mind the fact that there is really a specific number of very closely spaced discrete energy levels. In this way we can then apply the exclusion principle and allow just two electrons to occupy each of the states.

This probably all sounds rather abstract, so let us try to make these ideas more concrete by applying them to a specific case. Suppose we consider a tiny crystal of silicon containing one billion atoms. The previous argument shows that we will obtain a band of two billion closely spaced states, having a capacity to hold four billion electrons. This is precisely equal to the total number of valence electrons in the crystal, and so we conclude that the valence band is exactly filled. A similar analysis of any material in which only complete shells of electrons exist, in other words where the bonding is covalent, ionic or molecular, reveals that the valence band is always filled in these cases.

The situation is quite different for metals. If we consider a crystal containing a billion sodium atoms, we find that the lowest valence state in the sodium atom becomes a band which can hold two billion electrons. Since each sodium atom possesses only a single valence electron, this means that only half of the states in the valence band are occupied. From what we have said previously we expect the electrons to occupy the lowest energy states, leaving the higher energy states vacant, as shown in Figure 2.3. In this way we can define a characteristic energy

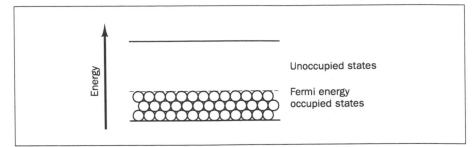

Figure 2.3 In a metal there are insufficient electrons to fill all the states in the valence band, consequently there are always some vacant states. The Fermi energy is a measure of the level to which the valence band is filled.

level, called the Fermi energy, which separates the occupied states from those which are unoccupied. For some other metals the situation is rather more complex than for sodium, but in all cases we find that the Fermi energy lies within the valence band, consequently there are always some vacant energy levels.

Let us now use these findings to examine once again the electrical conductivity of different materials. If we apply a voltage to a metal (for example, by connecting a battery across it) we can represent this as a tilt in the energy band, as shown in Figure 2.4(a), the positive terminal being at a lower energy than the negative end. (We will provide some justification for this in a moment.) The result of this tilt is that there are vacant energy levels towards the positive terminal which have lower energy than some of the occupied states near the negative end. Clearly, it is beneficial for the electrons to take advantage of these lower energy states, and so this picture is entirely consistent with that of electrons being attracted towards the positive end of the crystal.

What happens if our sample is an insulator? The energy band is tilted as for the metal (Figure 2.4(b)), but this time there are no vacant states towards the positive terminal since the entire band is full. Consequently, there can be no net movement of electrons in any direction. The situation is akin to that of a sweet jar crammed full of aniseed balls. We might expect to be able to move the sweets around by shaking the jar, but because they are so tightly packed no movement is allowed. In contrast, if we apply the same analogy to a metal, the jar is only partially filled, and so the sweets can be moved around with ease.

It seems then that we have obtained the same results as before. We can categorize materials either as conductors, which have only partially filled valence bands, or non-conductors having a full complement of electrons in the valence band. However, this is not the end of the story. Let us briefly return again to the hydrogen molecule that we introduced in the previous chapter. We found that the electrons in the molecule occupy a lower energy state than they do in

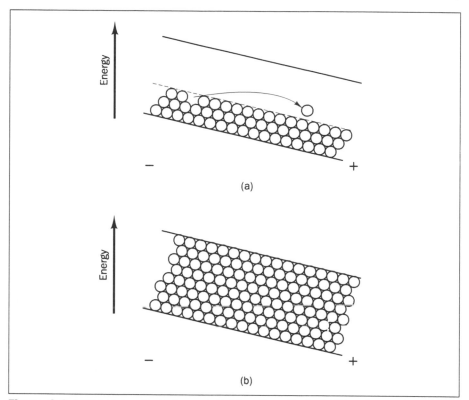

Figure 2.4 Applying a voltage to a crystal has the effect of tilting the energy bands. a) In a metal an electron can reach a lower energy state by moving towards the positive terminal. b) In an insulator no such movement is possible because the band is full.

a single atom of hydrogen. However, if we take a mathematical approach we find that there are actually two solutions to the hydrogen molecule problem. One is the lower energy state which we already know about, but the other corresponds to a state which has a higher energy than that of the isolated hydrogen atoms. The two states are shown schematically in Figure 2.5. Should we concern ourselves with this higher energy state? After all, both of the electrons can be accommodated quite comfortably in the lower state, and this is where we would expect them to reside. It seems then that we should be able to neglect the existence of this state, to dismiss it as a quirk of the mathematics, but let us pursue the consequences a stage further. Similar higher energy states occur not only in all other molecules, but also in solids where they interact to form a higher band of allowed energies. In non-metals there is generally a range of energies lying between these two bands in which there are no allowed energy levels. This is referred to as the forbidden energy gap, or simply the band gap, and is illustrated schematically in Figure 2.6.

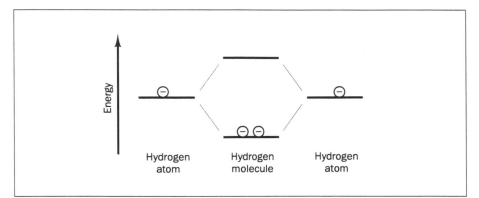

Figure 2.5 The energy levels in a hydrogen molecule relative to those in isolated hydrogen atoms.

Figure 2.6 In a non-metal there is a forbidden range of energies, the band gap, separating the valence and conduction bands.

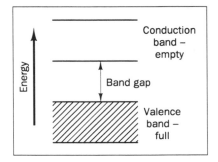

Once again we have to ask ourselves the same question as we did with the hydrogen molecule: what is the relevance of this higher band? Since all of the electrons can be accommodated in the valence band, we would expect the higher states to be vacant. However, let us suppose for one moment that an electron is somehow promoted to this higher band. It then has a multitude of vacant states from which to choose. This means that it can move about quite freely and therefore participate in electrical conduction. Consequently, this higher range of energies is called the conduction band.

How does an electron enter the conduction band? One of the ways it can achieve this is through gaining thermal energy. In order to explain what this means we need to consider how the energy of the electrons varies with temperature. To do this we will first of all consider a metal. In Figure 2.3 we suggested that the Fermi energy represents a distinct cut-off level—there are no valence electrons with energies greater than this, or vacant states with energies less than this. However, this is not strictly true. At any given temperature there will be a small proportion of valence electrons with energies greater than the Fermi energy (and accordingly there will be a similar number of vacant states

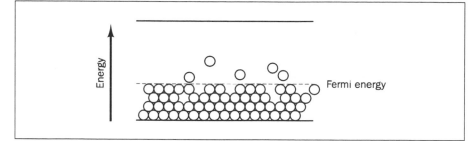

Figure 2.7 In a metal the effect of temperature is to raise a number of electrons up to states above the Fermi energy, leaving behind a corresponding number of vacant states below the Fermi energy.

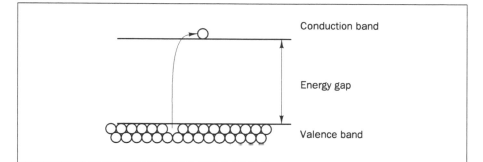

Figure 2.8 In an insulator or a semiconductor an electron can only move to a higher energy state if it gains sufficient energy to enter the conduction band. As a result, a vacant state is left behind in the valence band.

below the Fermi energy). This situation is shown schematically in Figure 2.7. As the temperature rises, the proportion of electrons above the Fermi energy increases. Consequently, we say that these electrons have gained some thermal energy.

The effect of temperature on the electrons in a metal is, therefore, quite straightforward. An electron just below the Fermi energy can gain an arbitrarily small amount of energy and move into one of the vacant levels above the Fermi energy, but for materials in which the valence band is full it is a different story. The presence of the forbidden energy gap means that it is not permissible for an electron to gain a small amount of energy, since this would require it to occupy a non-existent state in the gap. The only way an electron can move to a higher state is if it undergoes the process shown in Figure 2.8. In order to do this the electron must gain an amount of energy comparable with the band gap. For a crystal of silicon at room temperature the amount is about twenty times more than the average amount of thermal energy. Does this mean that we should not expect to find any conduction electrons in a crystal of silicon? On the contrary. Although the likelihood of any particular electron making the transition to the conduction band is extremely low, there are so many valence electrons that a large number manage to perform this feat. It is similar to saying that the chance of an expectant mother giving birth to quads is (thankfully) so remote that the possibility is unlikely to cross her mind. And yet if we examine the entire population of the Earth (some five and a half billion people) we will probably find about a million sets of quads.

The importance of all of this is that we can now see that the distinction between an electrical conductor and an insulator is not quite as clear cut as it first seemed. In non-metals there is always a possibility of finding some conduction electrons. However, since the number of conduction electrons is very low in comparison to a metal, the conductivity of the material is generally poor.

There is still one question which we have not yet resolved. How do we account for the difference in the ability of silicon and diamond to conduct electricity? In a cubic centimetre of silicon at room temperature we expect to find about ten billion conduction electrons. However, if we were fortunate enough to possess a pure diamond of this size we would be unlikely to find a single conduction electron. This is merely a result of the difference in the magnitude of the band gap in the two materials. The astonishing fact is that the band gap of diamond is only about five times larger than that of silicon. (We can provide some explanation for this by again considering the probabilities associated with multiple birth— although there are many instances of quads being born, there are no recorded cases of multiple births involving twenty children.) This extraordinary sensitivity to the size of the band gap results in an enormous difference in the ability of various solids to conduct electricity. For instance, the conductivity of a good metallic conductor, such as silver, is over a billion billion times better than that of the best insulators. Such a ratio is hard to appreciate, even for physicists who are traditionally used to dealing with large numbers. To try to visualize the magnitude of this difference we should consider that a similar figure relates the diameter of an atom to the distance from the Earth to the Moon, or the duration of one second to the age of the universe.

Having explored the differences between metals and insulators we will now concentrate on those materials which have a relatively small band gap. These are the materials known as semiconductors. The presence of a small band gap gives rise to some very interesting effects. For example, let us consider the effect of temperature on the electrical behaviour of a material. In metals it is well known that the resistance to an electrical current increases as the temperature rises. This is because the extra thermal energy causes the vibrations of the ions to increase, and consequently makes it more difficult for the electrons to pass through the structure. (A more detailed explanation of this effect is given in Chapter 6.) In a semiconductor the effect is reversed: the same temperature change allows the current to flow more easily. This anomalous result was noticed in the nineteenth century— indeed, Michael Faraday was possibly the first person to observe it, in 1833. However, it was nearly a hundred years before the origin of this effect was understood. We now know that it follows from the presence of the forbidden energy gap. At low temperatures, very few electrons can gain sufficient energy to enter the conduction band, but as the temperature rises we find that the number of electrons able to participate in conduction increases rapidly, producing a corresponding decrease in resistance. (In contrast, the number of conduction electrons in a metal remains constant with temperature.)

The most useful property of semiconductors, and the one which is of vital importance to the use of semiconductors in microelectronics, is that their conductivity can be controlled by adding tiny traces of impurity atoms, a process known as doping. Let us consider what would happen if we take a crystal of silicon and replace just one of the silicon atoms with a phosphorus atom. Since phosphorus has five valence electrons, it can share an electron with each of the four neighbouring silicon atoms, and still have one left over. The filled shells of electrons shield this extra electron from the full attractive force of the nucleus. In fact, they effectively cancel out all but one of the positive charges, so that we are left with an electron orbiting a single positive charge. This configuration is reminiscent of a hydrogen atom. However, since the phosphorus atom has obtained a full shell of eight electrons by bonding with the other silicon atoms, the extra electron is something of a nuisance, and so it is only very weakly bound to its parent atom. As a consequence, if we picture the electron orbiting around the remaining positive charge, we find that the radius of this orbit is typically twenty or so times larger than that of an electron in a hydrogen atom. This means that it encompasses a volume of the host silicon crystal containing over a thousand atoms. In this sense we cannot really picture it as belonging to any particular atom. Nevertheless, it is at least weakly confined to this region since if it moved away it would leave behind a positively charged ion.

Where does this extra electron fit into the energy band picture? There is no room for it in the valence band, since this would mean squeezing a third electron into one of the energy levels, contravening the exclusion principle. On the other hand, since it is constrained to a certain region, it cannot belong in the conduction band. The only answer is to put it in the band gap. This seems to be at odds with our previous statement that there are no allowed electron states in the forbidden energy gap. However, this criterion is strictly true only for a perfect crystal. By introducing an impurity the crystal is no longer perfect, and so it is quite reasonable for a state associated with this impurity to exist in the gap. As the electron is very weakly bound to its own nucleus, it requires only a small amount of energy to remove it to another part of the crystal where it can act as a conduction electron, and so the impurity state is placed very close

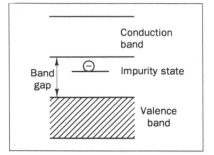

Figure 2.9 The extra electron associated with the impurity occupies a state just below the conduction band edge.

to the top of the band gap. This is shown in Figure 2.9. At room temperature the thermal energy will almost certainly be sufficient to liberate the electron, and so we can assume that each phosphorus atom introduces a conduction electron into the crystal.

An extremely low density of phosphorus atoms is required to alter significantly the properties of the material. In a typical case about one in every million silicon atoms will be replaced by a phosphorus atom. To appreciate how rare these impurities are, consider covering the entire surface of a full-sized football pitch with golf balls. If we assume that a silicon atom is represented by a white ball and a phosphorus atom by an orange ball, then on average we would find only two or three orange balls on the pitch. It is hard to believe that such a subtle change can have any noticeable effect, and yet this is quite sufficient for most conventional microelectronic applications.

The phosphorus atoms are referred to as donor impurities because they supply a conduction electron to the crystal. Silicon doped in this way is called n-type silicon, the 'n' indicating that there are negatively charged particles which are free to move through the crystal. It is important to notice, however, that the crystal itself is still electrically neutral. This is because the phosphorus nucleus has one more proton than a silicon nucleus, and so each conduction electron results in a positively charged ion being left elsewhere in the crystal. Of course, these ions occupy a fixed position in the lattice, so although there are both positively and negatively charged particles in the crystal, it is only the negative ones which participate in electrical conduction. These are referred to as the carriers. However, surprisingly, it is possible to have positively charged carriers in a semiconductor.

Suppose that instead of a phosphorus atom we introduce an atom of aluminium into the silicon crystal. We now appear to have a problem because aluminium has only three valence electrons and so it cannot fulfil all of the bonds which were broken when we removed the silicon atom. The three electrons form bonds with the neighbouring atoms, but this leaves one bond which can not be completed. We can see how this absence of an electron aids the conduction process by again considering the analogy with the sweets jar. If we take the filled jar of aniseed balls and remove one of the sweets, then when we shake the jar the small amount of extra space allows for some movement of the other balls. In a similar way, a space in the valence band allows the electrons to move about.

We have simplified matters slightly because in moving an electron into the incomplete bond we are adding an extra negative charge to the aluminium atom, and so a negatively charged ion is created. This requires a small amount of energy, and so strictly speaking the vacant state is in the band gap, having an energy just above the top of valence band. However, at room temperature a valence electron can easily gain the required energy to move up to this vacant state and so leave behind a 'hole' in the valence band.

Although this hole is simply a place where an electron isn't, it is usual to treat it as though it is a particle in its own right. We can see why this interpretation is beneficial by considering the following situation. Suppose we set out a group of chairs so that they form a row, one next to the other, down a gentle incline. We then find some people to sit on the chairs, one for each seat. Because of the slope, each person rests slightly on their neighbour in the seat below. We can imagine that this is analogous to the situation shown in Figure 2.4(b) if we assume that each person represents an electron in the energy band of a semiconductor and that a voltage is applied to the sample so that the lower end represents the positive terminal of the crystal. We now create a hole by removing one person from the middle of the row. This sets off a chain reaction. Since the person in the seat above now has no one to lean on, he shuffles down one space into the vacated seat. Similarly the next person up the row moves down into his seat, and so on. One way to interpret this is as a sequence of electrons each moving one step closer to the positive end of the crystal. However, from an alternative perspective we could describe the process as an empty seat moving up the slope. If we take a similar viewpoint to describe the motion of the hole, then we can see from Figure 2.10 that the hole is apparently attracted towards the negative end of the crystal. We therefore conclude that the hole has a positive charge, the magnitude of this charge being the same as that on an electron. We can now see that it makes perfect sense to visualize a particle with a positive charge moving through the crystal. On the other hand, if we insist that the hole is not really a particle then we have to come to terms with the fact that we have an electrical charge associated with an empty space.

A material doped in this way is called a p-type semiconductor since the carriers are positively charged holes. The impurity atoms in this case are known as acceptors since the change in conductivity arises as a result of the ability of these atoms to accept additional electrons.

Now that we have witnessed the important role played by holes in the valence band it is worth while re-examining the undoped case. We previously described how thermal energy can cause a valence electron to be excited into the conduction band. From the preceding argument and by re-examining Figure

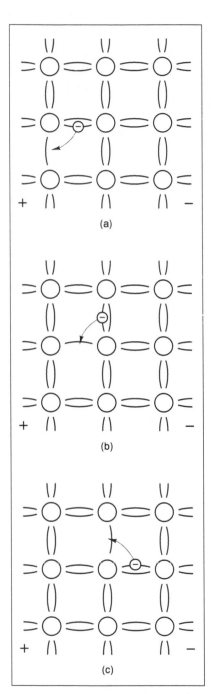

Figure 2.10 The successive movement of electrons through the valence band of a crystal illustrated by the sequence (a)–(c) is equivalent to a hole moving in the opposite direction.

2.8, we can now see that in the process a hole is simultaneously created in the valence band. We obtain two carriers for the price of one. It is therefore customary to talk about the creation of an electron-hole pair. At this stage it is also worth introducing two other technical terms— the undoped material which we assume to be completely free from impurities, is referred to as an intrinsic semiconductor, while material which has been intentionally doped is called an extrinsic semiconductor.

We can summarize these results as follows. In an intrinsic semiconductor the electrons and holes are always created in equal numbers, whereas if the material is doped we can choose whether the majority of the carriers are electrons or holes depending on whether donor or acceptor impurities are used. However, it is important to remember that both kinds of carrier are always present irrespective of the type of doping. For example, in a material doped with donors there will be some thermally created holes which we refer to as the minority carriers. This means that if an electric current is flowing through the material, then we really need to consider two separate currents, one consisting of a flow of negatively charged electrons, and the other of positively charged holes moving in the opposite direction. Since both the sign and the direction of the currents are different, we can see that the two add together to produce a total current.

What is the ratio of majority to minority carriers? We have seen that when doping a material we replace typically one in every million silicon atoms with an impurity. If we consider the dopant to be a donor, such as phosphorus, then each impurity should provide one conduction electron. Electrons and holes provided by thermal excitations are far more scarce. From the figures we gave earlier (viz. ten billion conduction electrons in a cubic centimetre of silicon consisting of a hundred thousand billion billion atoms) we expect to find only one pair of thermally produced carriers in a volume containing ten thousand billion silicon atoms. Once again we are faced with the difficulty of trying to comprehend numbers of this magnitude. Let us try to give some meaning to this by returning to the analogy of golf balls on a football pitch which we used previously to illustrate the density of dopant atoms. In this instance we would need enough golf balls to cover the surface of a small country, such as Luxembourg, to represent the number of silicon atoms corresponding to each minority carrier. Since the artificially introduced carriers outnumber the thermally produced ones by more than a million to one we might reasonably expect that the presence of the minority carriers can be ignored. However, as we shall see in the next two chapters, they play an important role in the operation of electronic devices.

We have seen that extremely low concentrations of impurities in semiconductors significantly alter the properties of these materials. It is hardly surprising then that much of the early work on semiconductors was plagued with problems of controlling the levels of accidental impurities already present in the material.

At the time this posed a serious technological problem. It is easy to see why. The effects we have discussed require intentional doping densities of only one atom in every million silicon atoms. In order to observe these effects the concentration of accidental impurities must be far smaller. Otherwise, we may find, for example, that the presence of these impurities produces a large number of holes in what is designed to be an n-type material. It is imperative, therefore, that the density of accidental impurities is reduced to as low a level as possible, and certainly much lower than the level produced by intentional doping.

This is rather a tall order since the best that can be achieved by chemical means is typically one impurity in every million atoms. How can the impurity level be reduced further? There are several ways, but most employ in some form a technique called zone refining. The basic principle behind zone refining is that if we form a crystal from an impure solution, the most pure material will crystallize at the highest temperature. The starting-point is a column of the material which has already been purified as far as possible by chemical means. Using an electrical heater around the column we concentrate the heat so that just a thin disc of the column is melted at any one time. As the heater moves slowly down the column the molten disc moves with it. Accordingly, some of the previously melted material above this re-crystallizes. Since it is the purest material which crystallizes first, the impurities tend to move downwards with the molten section of the column. In this way we can redistribute the impurities so that the greatest concentration accumulate at the bottom of the column, leaving a relatively pure material at the top. By repeating this process many times it is possible routinely to produce large samples of semiconductors with less than one impurity in every ten billion atoms. It is a sobering thought that without this degree of control over the purity of semiconductors the fabrication of microelectronic circuits would not be possible.

We have spent most of this chapter considering how to create free carriers in semiconductors, but once they are there they do not exist forever. This is obvious if we consider that the excitation of electrons into the conduction band is a thermal process which goes on continually. If the electrons did not return to the valence band, the number of conduction electrons in the sample would steadily increase with time. In addition, we know that electrons tend to occupy the lowest energy states available. If there are vacant states in the valence band then a conduction electron will find these states far more attractive than its present situation. Thus, we find that after a short period of time, typically measured in millionths of a second, a conduction electron will recombine with a hole in the valence band, effectively annihilating both carriers. (Of course, we are left with a valence electron, but this can not take part in electrical conduction.) In any system which is in equilibrium the rate at which electron–hole pairs are created must be equal to the rate at which they are destroyed. In this way the concentrations of conduction electrons and holes are kept constant.

One question remains to be answered: since a valence electron has less energy than a conduction electron, what happens to the excess energy when they recombine? We have already answered this question in the previous chapter when considering electrons moving between energy states in atoms. In most cases the energy is emitted as a light wave. Since the electrons tend to occupy the lowest energy levels in the conduction band, and the holes generally are at the top of the valence band, the wave length of the light emitted corresponds to the band gap energy.

We have now examined some of the principal properties of semiconductors. In particular, we have found that the conductivity of these materials can be altered substantially by introducing low levels of impurities. In the next chapter we will consider how this can be used in a device and also explore some applications of the optical properties of these materials.

3

p-n Junctions
How They Work and What You Can Do With Them

S O far we have considered the properties of various types of solids, and, in the case of semiconductors, seen how we can alter these properties. In order to make a device we need to combine two or more materials with different characteristics. We will start with the p-n junction. This is the simplest semiconductor device in terms of structure; however, it is by no means simple to understand. Several effects must be taken into account, but we will work through these slowly, describing each in turn before looking at some of the applications of the device.

First of all we should clear up one common misunderstanding. The name p-n junction is misleading in that it implies that we start off with a crystal of p-type semiconductor and another of n-type semiconductor and somehow join the two together. If we did this we would not end up with a single crystal. Instead we would have two crystalline regions separated by a disorganized jumble of atoms at the interface. In addition, the surfaces of the crystals will be contaminated with a large number of impurities. If we fuse the two together then these impurities will be incorporated into the middle of the new structure. Such effects would completely destroy the phenomena that we are about to describe in this chapter. It is therefore essential that we start off with one single crystal of semiconductor. The 'junction' is then created by selectively introducing donor atoms into one half of the sample and acceptors into the other half.

Suppose that we have now fabricated such a structure. We can assume that initially there is a majority of conduction electrons on one side and of holes on the other. Will it remain this way? It seems unlikely. Having virtually all of the conduction electrons concentrated in one half of the crystal and the holes in the other half represents a highly unnatural situation. It is like having a swimming pool, one end of which contains salt water and the other end fresh water. Even if we start off with a freshwater pool and add salt only at one end, within a short while the whole of the pool will have the same average salt concentration. This is a result of the random motion of the water and salt molecules, a process known as diffusion. We could demonstrate the same effect if we blindfolded a group of people and placed them in the middle of a field. As time progressed we would observe the group spreading out as each person followed a random course, until the participants were distributed evenly across the field. Although the individuals would continue to move, it would be highly unusual for them to re-form into any concentrated group. In fact, this tendency of any concentrated collection of particles to spread out and produce a uniform distribution is so compelling that it is usual to think of diffusion as a force which pushes the particles from the densely populated areas into the unoccupied regions.

Let us apply this principle of diffusion to the conduction electrons and holes in our p-n junction. With no other considerations we would predict that the conduction electrons would spread out across the p-type region, and similarly the holes would redistribute themselves over the n-type region. However, it is not quite so straightforward. We have neglected to take into account the ions which initially gave rise to the concentrations of electrons and holes. Diffusion has no effect on the positions of the ions because they are fixed in the crystal lattice. Consequently, if the carriers were spread out evenly across the whole of the crystal, we would still be left with all the negatively charged ions on one side and all the positively charged ions on the other. Such an imbalance of charge is also unnatural, so we would expect something to happen before this stage is reached. We therefore have to consider two opposing forces, one being due to diffusion which causes the electrons and holes to spread out, and the other being the tendency towards the neutral distribution of electrical charge. In order to see how these forces reach equilibrium we will follow the process step by step.

Let us again start from a crystal in which both sides are electrically neutral, the conduction electrons balancing out the positive ions on the n-type side and the holes balancing out the negative ions on the p-type side. We will now consider the effect of a single conduction electron diffusing into the p-type region. This results in a very slight imbalance of charge. On the p-type side there is exactly one more electron than there are protons, whilst an uncompensated positive ion has been left behind on the n-type side, as shown in Figure 3.1. The redistribution of charge has two effects. Firstly, the electron feels an attractive force from the now positively charged n-type region of the crystal, which means that it is unlikely to stray far from the junction. The second effect is that the slight negative charge on the p-type side of the junction makes it harder for a second electron to enter this region. Already we can begin to see how the transfer of charge across the junction opposes the effects of diffusion. As more electrons cross into the p-type region the same principles apply. The electrons do not move far into the adjacent layer, and so the charge accumulates at the junction. Finally, the situation arises where the repulsive force from this accumulated charge balances the diffusive force tending to move electrons in the opposite direction.

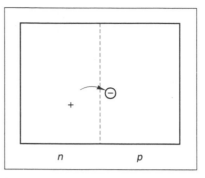

Figure 3.1 The effect of an electron crossing into the p-type region is to leave behind a net positive charge in the n-type region.

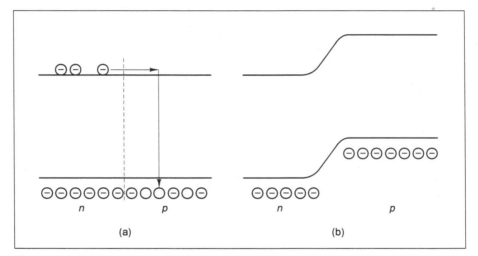

Figure 3.2 As electrons cross into the p-type side and recombine with a hole (a), the energy step at the junction becomes larger (b).

Another way to look at this is in terms of the energy of the electrons. The first electron senses no difference in energy as it moves into the p-type region. As the second electron crosses the junction, it now has to struggle against the repulsive force from the first electron, and so it must use a small amount of its energy to overcome this force. This means that a conduction electron in the very lowest energy level on the n-type side no longer has sufficient energy to move across the junction. Since such an electron is not permitted to enter the p-type side we can say that the lowest conduction band state on this side must be slightly higher in energy than that on the n-type side. This produces a step in the edge of the conduction band, as shown in Figure 3.2. The effect is very small when we are talking about a single electron, but as successive electrons are transferred to the p-type side, the magnitude of the step increases, meaning that only those electrons in the higher energy states continue to be able to cross the junction.

We might eventually expect to reach a situation where the difference in energy is so large that none of the conduction electrons on the n-type side has sufficient energy to overcome the step. At this point there would be no further flow of charge. Actually this is not quite correct since we have neglected to consider the minority carriers. As we have seen in the previous chapter, there will always be a small concentration of conduction electrons in a p-type semiconductor. These will be tempted to move in the opposite direction in order to take advantage of the lower energy states on the n-type side. This means that there will be electrons moving in both directions across the junction. Those flowing from n to p constitute what is called the diffusion current, because it is the diffusion process that causes these electrons to cross the junction. The flow of minority carriers in the opposite direction is described as the drift current. The equilibrium condition is achieved when these two currents are equal, i.e. when the net current is zero. It is important to remember, however, that even though there is no net current, electrons still continue to flow across the junction—it is just that the rates of flow in each direction are equal. We should therefore refer to this condition as a dynamic equilibrium.

These two descriptions of the p-n junction are complementary, and we can use whichever seems most appropriate in a particular situation. A slightly less accurate picture, but one which illustrates the ideas without having to worry about the physics, is to think of a p-n junction as being a lawless border region between two neighbouring countries. The two tribes of people on either side make constant raids on each other's land and there is much bloodshed. However, a ruthless form of justice operates from which there is no escape—in every case the one who kills always ends up being killed. One important point is that it is not a war between the two countries. The skirmishes are confined to the border region only. As far as the rest of the population of these countries is concerned, life goes on more or less as normal.

The same is true of the p-n junction. As we mentioned previously, an electron which crosses over to the p-type region does not tend to venture far into this layer as it is attracted towards the net positive charge left behind on the other side of the junction. After a short while it is likely to recombine with one of the holes, so removing one carrier of each type. As more electrons cross the junction this process of mutual annihilation gradually depletes the number of carriers in the vicinity of the interface, and as equilibrium is achieved we find that there are very few carriers of either type within this region. This is known as the depletion layer and typically extends a few microns (a micron being a thousandth of a millimetre) either side of the junction. Beyond this region the crystal is largely unaffected by the presence of the interface.

So far we have portrayed a very one-sided picture of the p-n junction. The description has been entirely in terms of electrons moving across the junction, but of course we should also include the possibility of holes travelling from one side to the other. There is nothing to stop us from following the same line of reasoning for holes as we did with the electrons. We would notice a build-up of positive charge on the n-type side of the junction as the holes diffused into this layer, with a corresponding negative charge, due to the uncompensated acceptor ions, forming on the p-type side. This would lead to the creation of an energy step in the valence band, as illustrated in Figure 3.2(b). At first sight it appears as though the energy step is in the wrong direction. As holes move across the junction, the top valence band state on the n-type side becomes lower in energy than that on the p-type side. This would seem to encourage further holes to cross into the n-type side. However, we have neglected to take into account the perverse nature of holes which is that they tend to favour the higher energy states (we will consider this unusual behaviour in more detail in Chapter 6). We therefore find that the results are very similar to those we obtained when considering electrons. As the holes diffuse across the junction it becomes increasingly difficult for others to follow.

A simple argument will serve to show that it does not really matter whether we consider the movement of carriers to be due to electrons or holes. If we imagine a hole travelling into the n-type layer and then recombining with a conduction electron, the result is that the n-type layer gains a positive charge (at the expense of losing one of its conduction electrons), whilst the p-type layer gains a negative charge (at the expense of losing a hole). Alternatively, if we consider a conduction electron moving in the opposite direction and recombining with a hole the result is exactly the same. Consequently, in this and further discussions we are justified in confining our attention to just one type of carrier.

*

The above describes how a p-n junction arrives at an equilibrium situation. However, if we want to make use of it as a device we need to examine what happens when a voltage is applied. Suppose that the n-type side is attached to the negative terminal of a battery, and the p-type side to the positive terminal. As initially (i.e., in equilibrium) there is a net positive charge on the n-type side of the junction, the electrons will flow into this side of the crystal from the battery in an attempt to compensate for this charge. It is slightly harder to imagine, but we can also think of holes flowing into the opposite side of the crystal from the positive terminal of the battery. (An easier way of dealing with this is to say that it results in electrons flowing out of the crystal to the positive terminal of the battery.) How does this affect the flow of electrons across the junction?

We found previously that the tendency of electrons to diffuse into the p-type region is opposed by two forces: repulsion due to the accumulation of negative charge on the p-type side, and attraction arising from the positive ions left behind

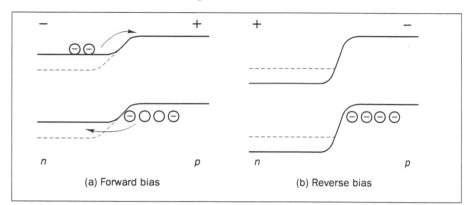

(a) Forward bias   (b) Reverse bias

Figure 3.3 (a) Under forward bias the energy step is reduced, allowing electrons and holes to cross the junction. (b) With reverse bias the energy step is increased, and very few carriers cross the junction.

on the n-type side. As both of these effects are reduced by the applied voltage, the electrons can now move much more easily into the p-type region. We can interpret this by saying that the difference in energy between the two sides of the junction has become smaller. The result is shown in Figure 3.3(a). We therefore find that there is a net movement of electrons across the junction. In other words the diffusion current now exceeds the drift current.

What happens to the conduction electrons once they reach the p-type side of the junction? They no longer congregate around the junction since they are now attracted towards the positive terminal of the battery. Not all of them make it this far because they are now moving through enemy territory, in a region where there is a comparatively large population of holes. Consequently, many of them are killed off, by recombining with holes, before they make it to the positive terminal. They do not die in vain, however, since the battery is under an obligation to maintain the number of holes on the p-type side (to balance out the negative charge of the acceptor ions). Accordingly, more holes are injected into this region (i.e. electrons flow out) in order to compensate those lost by recombination. Since electrons flow in at the n-type side, and out through the opposite side, we say that a current flows through the device. Moreover, if we increase the voltage by a small amount, the energy step is correspondingly reduced, and far more electrons are able to cross the junction. We therefore find that the current flowing through the device increases rapidly as the voltage is altered.

The situation is quite different if the electrical connections are reversed. Attaching the negative terminal of the battery to the p-type side means that we are trying to introduce electrons into a region which is already negatively charged. In doing so we increase the energy step (see Figure 3.3(b)), and so reduce the rate at which electrons diffuse across the junction. On the other hand, the drift current which flows in the opposite direction is insensitive to the magnitude of the energy step since it consists of thermally produced conduction electrons from the p-type layer finding their way into lower energy states on the n-type side. We therefore find that the drift current exceeds the diffusion current. In other words there is now a net flow of electrons from the p-type to the n-type side. However, as we are dealing with minority carriers, the number of electrons involved is very small.

Consequently, we find that the current flow through a p-n junction depends critically on the polarity of the electrical connections. In one case, where the n-type layer is made negative and the p-type layer is made positive, the current of electrons flows through the device like a mighty river. This situation is referred to as forward bias. In the other case, reverse bias, the flow is reduced to a tiny trickle. This enormous difference is due to the fact that in the former case we are dealing with the movement of majority carriers, while in the latter it is only the minority carriers which take part. Already we are beginning to see how the ability to dope a semiconductor, and therefore control the number and type of carriers, can produce novel results. We will take this a stage further in the next chapter when we examine the transistor, but first we will take a look at some applications of the p-n junction.

*

The asymmetric behaviour of the p-n junction is exploited in a rectifier, the purpose of which is to convert an alternating current into a direct current. When used in this context the device is referred to as a diode. The principle of operation of the diode depends directly on the properties that we have just described. If an alternating voltage is applied, then for half of the cycle the input to the n-type side is negative and a current flows, and for the other half of the cycle the input is positive and virtually no current flows. Figure 3.4 shows that although the output is not constant, it is at least always of the same polarity. By using a more complex arrangement of diodes it is possible to obtain a direct current which is approximately constant. Such a rectifier is a vital part of any mains-powered computer since the alternating mains current must be converted into a direct current in order to power the electronic components.

Many other applications of p-n junctions make use of their optical properties. We have seen that light can be produced in a semiconductor when conduction electrons recombine with holes. In general, these are random events which occur frequently throughout the crystal. Although such processes are going on all the time, each event produces such a tiny pulse of light that we are unlikely to see them. This is in accord with observation: lumps of semiconductor do not emit light! If we wish to make use of this property and produce observable intensities of light, then somehow we must increase the frequency with which these events

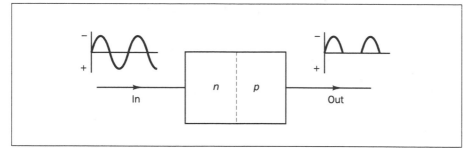

Figure 3.4 Since a p-n junction allows current to flow in only one direction, it acts as a rectifier when an alternating current is applied.

occur. The problem is that we require the presence of both an electron and a hole in order to generate the pulse of light. Even if we dope the material, the thermal production of electron–hole pairs is the only mechanism for maintaining an equilibrium level of the minority carriers.

The solution is to have both types of doping present in close proximity. Of course, this is precisely what we obtain with a p-n junction. In addition, we ensure that most of the recombination events are concentrated in a narrow region either side of the junction, and not at some arbitrary position in the crystal. If we apply a forward bias to the junction, this provides a constant source of conduction electrons and holes which move into the region from opposite sides. As each pair recombine they are replaced by the battery. By varying the applied voltage we can control the rate at which the electrons and holes recombine, and thereby control the intensity of light produced. These devices go by the name LED, or light-emitting diode. They are now a familiar sight as display indicators in many electrical products.

As we discovered in Chapter 1, the wavelength, or colour, of the light depends upon the amount of energy the conduction electron must give away in order to recombine with the hole. In general, this is determined by the magnitude of the band gap of the semiconductor. Consequently, the use of different types of semiconducting materials results in various colours of light. Red and green light are most easily achieved, but more recently materials have been developed which produce blue light. There is, however, a restriction governing the types of semiconductor materials which can be used to produce these devices. This is because the process by which electrons and holes recombine is very inefficient in certain semiconductors such as silicon. (We will examine the reasons for this in Chapter 8.) Instead, the preferred materials for light-emitting diodes and semiconductor lasers (which we will deal with in a moment) are gallium arsenide and related compounds.

If electrons and holes can recombine to produce light, then it is a reasonable assumption that if we shine light of the correct wavelength on to a semiconductor we can create an electron–hole pair. This turns out to be correct: the energy from the light wave is transferred to a valence electron so that it is promoted to the conduction band. The effect has several applications. For instance, if we illuminate a semiconductor, the extra conduction electrons and holes which are created increase the conductivity of the sample, a phenomenon known as photoconductivity. This effect can be used to detect the presence of light. Longer wavelength signals in the infrared region of the spectrum can also be detected, providing that the band gap of the semiconductor is small enough. One way of producing such a material is to form an alloy of the compounds cadmium telluride and mercury telluride. Cadmium telluride is a semiconductor with an energy gap somewhat larger than that of silicon; whereas mercury telluride is a semi-metal, which means that, as in a metal, there is no forbidden energy region between the valence and conduction bands, although the material has many other properties which are characteristic of a semiconductor. The resultant alloy, cadmium mercury telluride, or CMT for short, can have a band gap energy anywhere in the range between zero and the band gap of cadmium telluride, depending on the relative proportions of the two compounds. However, it is near the lowest extreme of this range, where the energy of the band gap corresponds to the radiation emitted by a warm human body, that these materials have found most applications. In this case they are used in 'night sights' and other similar equipment.

A slight variation on this idea entails shining light on to a p-n junction. Conduction electrons created in the p-type layer near to the junction see a far more attractive (i.e. lower) energy level available in the n-type layer, and so tend to move across the junction. Similarly, holes created in the n-type layer tend to move in the opposite direction. This alters the distribution of charge at the interface. In fact, it resembles the situation when we apply a forward bias, except, of course, that there is no external voltage applied to the system. Consequently, when used in this manner the p-n junction acts as a source of voltage, just like a battery, and so we can use it to power an electrical circuit. This is the principle behind the solar cell used in many applications from hand-held calculators to the solar panels that provide the power for satellites and space stations.

There is one other related device that is of great importance: the semiconductor diode laser. The rapidly growing communications industry based on optical fibres requires compact lasers with low power requirements—precisely the characteristics that diode lasers can offer. However, before we look at the diode laser we need to be quite sure about what a laser does. A laser is a device which produces a very intense beam of light with a single wavelength. The other key feature is that the light produced by a laser is coherent. What does this mean?

Let us answer this by first considering some examples which are not coherent. For instance, in a light-emitting diode the light is produced as the result of a sequence of unconnected events. What we mean by this is that the recombination of a conduction electron and a hole is not influenced by the action of any of the neighbouring electrons. Each pulse of light is generated at a random moment in time, and although all the waves are of similar wavelength, they are not all at the same stage at the same time. The process is called spontaneous emission, and it results in incoherent light.

A more visual example of incoherence can be obtained with water waves in a bath tub. It is easy to make a wave in a shallow bath simply by moving your hand a short distance through the water. Suppose we make two such waves in quick succession. The first wave travels down the bath, is reflected by the end of the bath and starts moving back towards us. After a brief moment it interacts with the second wave travelling in the opposite direction. What happens? The waves may move past each other unhindered and continue on their way, or they may collide in a big splash leaving virtually no wave at all. Which of these possibilities actually occurs depends on what stage each of the waves is at, in other words, on the phase of the waves. One thing we can be certain of is that a large number of randomly produced waves will, on average, cancel each other out.

Coherent waves are quite different since each wave is at exactly the same phase as all the others. Such waves are very simple to make in the bath. We make one wave and allow it to travel down the bath, reflect back up the bath again, and then reflect off the other end. As it comes back past us we make a similar wave directly on top of the first. We now have a much larger wave moving down the bath. If we repeat this a third time we will probably find that the water sloshes right out over the end of the bath because the amplitude of the wave has become so large. It may make a mess, but it is a clear demonstration of the way in which we can dramatically amplify a wave by combining several waves which are in phase.

Achieving coherent water waves is simple enough because we can actually see the shape of the wave. With light waves it is rather more difficult. Even if we could see the light waves, how could we ensure that the conduction electron would recombine at just the right moment? Fortunately, nature has already supplied us with the answer. Under certain circumstances, the process of one conduction electron recombining with a hole may trigger other electrons to follow suit. Let us picture a light wave produced by one such event passing by a conduction electron which is poised ready to jump down into the valence band. The passing light wave is just the incentive that the electron needs, and so it recombines and produces an identical light wave with exactly the same phase as the first one. This process is called stimulated emission, since it is the presence of the light wave which stimulates the second electron to recombine. We can see from Figure 3.5 that as further similar events are induced, the light wave grows in intensity. If we polish two opposing faces of the crystal so that they

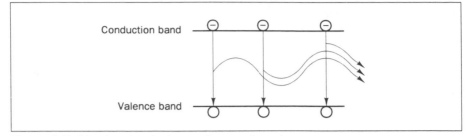

Figure 3.5 Stimulated emission produces coherent waves.

act as mirrors, then most of the light generated in the p-n junction is reflected backwards and forwards through the crystal, being amplified on each pass, just like the water wave in the bath tub. In this way it is possible to produce a very intense beam of coherent light. In fact, the name laser is really a description of these processes: it is an acronym for light amplification by stimulated emission of radiation. The one remaining difficulty is: how do we obtain stimulated emission rather than spontaneous emission?

Let us consider the following scenario. An electron and a hole recombine, and in the process produce a short pulse of light. This light wave travels through the crystal. However, a moment later it is likely that the light wave will once again disappear as its energy is used to raise another valence electron into the conduction band. One way to reduce the chances of the light wave being re-absorbed is to construct the device so that the junction is very close to the surface from which we wish the light to be emitted. This is a usual feature of light-emitting diodes, but such an arrangement is of no use for a laser since we want the light wave to travel many times through the crystal in order that it can stimulate other conduction electrons to recombine. The root of the problem is that in most cases it is far more likely for a light wave to create an electron–hole pair than it is for it to stimulate the production of another light wave. It is simply a question of numbers. There are a large number of valence electrons in a suitable position to be excited into the conduction band, but there are very few conduction electrons available as potential candidates to recombine. Somehow we need to reverse the odds. To do this the conduction electrons and holes must outnumber the electrons at the top of the valence band, a condition know as a population inversion. The two situations are shown in Figure 3.6. How do we produce a configuration such as that in (b)? Thermal energy can be used to excite electrons into the conduction band, but the amount of energy required is far greater than the average thermal energy and so only a very small proportion of the electrons are in the conduction band at any given time. Consequently, it is highly improbable that we would achieve a population inversion by chance.

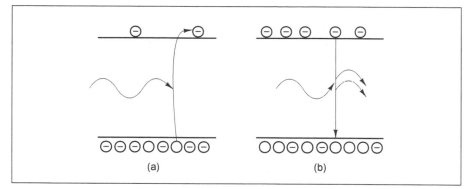

Figure 3.6 (a) Under normal conditions a light wave is absorbed, creating an electron–hole pair. (b) With population inversion a light wave is most likely to produce stimulated emission.

Instead we must somehow artificially create this inverted population in a particular region of the crystal.

The key to achieving this is to use very high densities of doping. Typically one in every thousand atoms in the crystal must be replaced with a suitable dopant atom. If we form a p-n junction between two such heavily doped layers, then large numbers of conduction electrons will transfer to the p-type side (and similarly holes will move in the opposite direction) under the influence of diffusion. This creates a population inversion around the junction region. Of course, this situation will only persist for a fraction of a second, by which time a substantial proportion of the conduction electrons and holes will have recombined. If we wish to maintain the population inversion then it is necessary to apply a large forward bias to the junction in order to constantly replenish the supplies of conduction electrons and holes. Since a large current flows, this generates an enormous amount of heat. Great efforts are required to cool the device to stop it from burning up, and consequently it is only possible to operate the laser in short bursts, or pulses. Nevertheless, it seems quite amazing that a tiny piece of semiconductor can be the source of such an intense beam of light. We shall see later on, in Chapter 7, that further improvements are possible—a slightly different approach can drastically reduce the power consumption, making it possible to build continuously operating hand-held lasers which require no special cooling apparatus.

We have dealt principally with the optical properties of semiconductors and p-n junctions in this chapter. These have found many technological applications in a wide variety of situations. For instance, light-emitting diodes are used in virtually every electrical gadget, where they are cheaper, smaller and many times more reliable and robust than using small light bulbs. The increasing use of optical fibres has dramatically altered the world of communications, and semiconductor lasers and detectors play a vital role in converting an electrical signal into an optical one, and in decoding the signal at the other end of the fibre. However, the greatest impact of semiconductors has been through the development of the transistor. We will take a look at this remarkable device in the next chapter.

4

A Logical Decision
Using the Transistor as an Electronic Switch

THE idea of creating a machine which can automatically perform mathematical calculations has been around for much longer than most people imagine. The earliest attempts to build a calculating machine are usually credited to Blaise Pascal and date from the mid-seventeenth century. These machines were very limited in their abilities, being capable only of addition and subtraction (the latter being performed after making a small adjustment to the internal workings). Although the first examples were rather unreliable, many improved versions appeared throughout the following two hundred years. Some of these later models were also capable of performing multiplication and division using a special gear wheel designed by Gottfried Leibniz.

The next major advance came in the 1830s. The English mathematician and inventor Charles Babbage had far more ambitious ideas. He proposed a programmable machine, called the Analytical Engine, which would be controlled by punched cards. (Similar cards were already in use to determine the patterns produced by automated Jacquard looms.) Although Babbage never succeeded in building a working model, the Analytical Engine is generally considered to be the original antecedent of the general-purpose digital computer. Indeed, many of the concepts which he proposed, such as the idea of a stored program, were actually redeveloped a hundred years later by researchers who were unaware of Babbage's earlier work.

If the ideas proposed by Babbage were sound, why is it that he was not able to build a working Analytical Engine? One problem was that the machine consisted of a mass of gear wheels. The tolerances to which these gears had to be produced and assembled were simply beyond the manufacturing capabilities of the time. However, the underlying reason was that the machine dealt with numbers in base ten. Although this might seem to be the obvious choice, since it is the number system that we are all familiar with, it is not the best suited to mechanical calculations. The reasons are simple. To represent a number in base ten we require ten different digits, 0 to 9. At each stage of the calculation it is

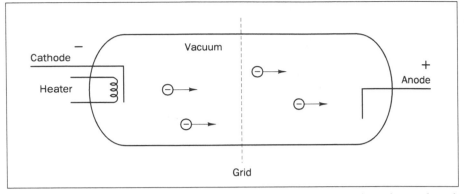

Figure 4.1 A schematic representation of the flow of electrons through a valve. A voltage applied to the grid controls the flow of current through the device.

necessary to be able to determine the value of each digit with absolute certainty. Since there are ten possibilities from which to choose, an extreme degree of mechanical precision is required. Bearing these considerations in mind it is clear (with the benefit of hindsight) that the ideal number system is binary, or base two. In this case there are only two digits, 0 and 1, and so at any stage we need only to be able to distinguish between one of two different states. In other words, all that is needed is a switch which can be in one of two positions, either ON or OFF. Consequently, the use of binary notation allows for much greater tolerances in the individual components.

It was not until the 1930s that the importance of using the binary representation of numbers in a machine was appreciated. By this time there were two suitable technologies available. One involved using electro-mechanical switches, in which the application of an electrical signal causes a small lever to move and so physically redirect the flow of current. Such devices were already in use in telephone exchanges where they had proved to be both cheap and reliable.

The alternative was the vacuum tube or valve. In its simplest form a valve consists of an evacuated glass envelope, like a long thin light bulb, containing an electrode at each end. A schematic diagram of such a system is shown in Figure 4.1. By heating the negative electrode (cathode), electrons are emitted. These travel along the length of the tube, attracted by the positive electrode (anode) at the other end. Consequently, a current flows through the tube. In order to make a switching device a third electrode is added part way along the tube. We will not go into great detail as to how the device works, but merely note that by applying a voltage to this electrode the electrons can be prohibited from reaching the anode. We can therefore determine whether or not a current flows through the device simply by applying a voltage to this third electrode. We describe the valve as an electronic switch since it contains no moving parts and yet has two distinct states of operation.

Electronic computers constructed using valves began to appear in the early 1940s. The first such machine was probably the British-built Colossus, containing about two and a half thousand valves. It was developed by a team of scientists working under top security, because the purpose of the machine was to crack the ciphers that the Nazis were using to send coded messages. Within a few years the Americans had built a vastly more powerful computer, also for military applications. Called ENIAC, standing for electronic numeric integrator and calculator, it was a monster of a machine containing nineteen thousand valves.

The valves were expensive, relatively large, and consumed considerable amounts of power. (ENIAC required about the same quantity of electrical power as one hundred and fifty one-bar electric fires.) However, the principal problem was with the reliability of the tubes. They suffered vacuum leaks which caused the devices to malfunction, and repeated heating of the cathode would eventually cause the valve to blow, just like a light bulb. Let us suppose that under these conditions the lifetime of an individual valve was approximately three thousand hours, or about four months of continuous operation. This sounds quite adequate. However, with nineteen thousand valves it meant that on average one would blow every ten minutes. Consequently, it was virtually a full-time job to find and replace the blown valves. Fortunately, this dinosaur was condemned to an early death by the invention of the semiconductor transistor a few years later.

The transistor has many advantages over the valve. It is far smaller, uses tiny amounts of power, can be made at exceptionally low cost by mass production methods, and has a virtually infinite lifetime. As a result, it has not only taken over the role of the valve in television sets and computers, but it has also allowed innumerable applications in a wide range of environments where a valve could never be used. The transistor performs vital functions in virtually every piece of electrical equipment, and yet to a large proportion of people it remains a rather mysterious object. So let us begin by looking at what a transistor does. In many applications the transistor functions as a switch. We can think of it as being like a light switch, having two distinct states. The switch is either in the OFF position, in which case no current flows, or it is in the ON position and the current flows freely. As we discussed above, this is the principal use of transistors in computers and the one we will concentrate on in this chapter, but a transistor can also be used as an amplifier. In this mode a small input signal is faithfully magnified, just like the operation of an amplifier in a hi-fi system. There are basically two distinct types of transistor and we will examine in turn how each of these operates. Both can be used either in the switching or amplifying modes.

We will begin with the FET, or field effect transistor, which in many ways is the simplest to understand. There are many variants on this theme, but virtually all are recognizable by the inclusion of the acronym FET at the end of the name.

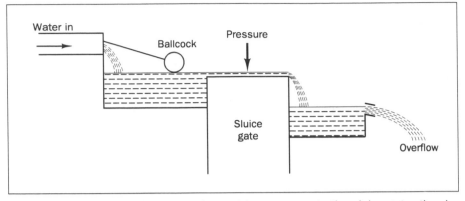

Figure 4.2 The water FET operates by applying pressure to the sluice gate, thereby allowing water to flow through the system.

We will concentrate on a structure called the MOSFET, which for many years has been the most commonly used type of FET. The initials MOS stand for metal oxide semiconductor, the three essential components of this transistor. However, first we will consider a more tangible model, one which you won't find inside any electrical gadget —a water FET.

The basic elements of a water FET are shown in Figure 4.2. It consists of two tanks of water, one at a higher level than the other. The upper tank has a ballcock arrangement to control the level of water in the tank, while the lower one has an overflow pipe for any excess water to run out. The water is prevented from flowing into the lower tank by the presence of a sluice gate. To operate the sluice gate we simply press down on its upper surface. When the top of the gate passes below the level of the water in the upper tank, the water flows across the gate, into the lower tank and out through the overflow. As the water level in the top tank begins to drop, the ball cock allows more water in to keep the level constant. Consequently, a stream of water flows through the system all the time that the gate is depressed, and ceases to flow when the pressure is released. We can therefore think of the system as having two states, which we can denote as ON and OFF, corresponding to whether or not water flows through the system.

Let us now consider a real transistor, the MOSFET. It consists of a p-type semiconductor with two n-type regions, one at either end, as shown in Figure 4.3. We can think of this as two p-n junctions placed back-to-back. If a voltage is applied to the two ends of the crystal, virtually no current will flow, irrespective of the polarity of the terminals, since one of the junctions will always be reverse biased. This does not seem to be of much use. However, the ability of the transistor to act as a switch is dependent on a third electrical contact. On the top surface of the structure, above the p-type region, is a thin layer of insulator. This is very easy to create when using silicon since the natural oxide of silicon

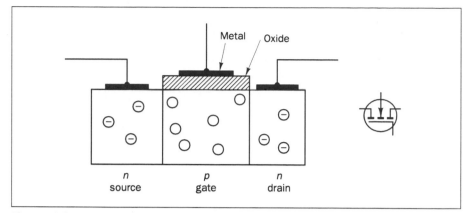

Figure 4.3 The basic elements of a MOSFET. The circuit symbol is also shown, the direction of the arrow indicating that the gate region is a p-type semiconductor.

forms an excellent insulator. The third metallic contact, the gate electrode, is made to the surface of this insulator. The central region therefore consists of a coating of metal and a layer of oxide above the semiconductor substrate—the three ingredients from which the name MOSFET is derived. By applying a positive voltage to the central electrode it is possible to change the behaviour of the underlying semiconductor. This happens because the positive charge on the gate electrode repels the holes at the top surface of the p-type layer, and pushes them deeper into the material. In addition, we know that there will be a small number of thermally created conduction electrons (minority carriers) in the p-type region. These will be attracted by the positive charge on the gate electrode. However, they cannot flow out through the gate contact, since to do so they would have to pass through the insulator, so instead they collect at the surface of the semiconductor. They are unlikely to recombine with holes, because most of the holes have been pushed away from the surface. Consequently, we find that the conduction electrons become the majority carriers in this region of the p-type material. In other words, by applying a positive voltage to the gate we have created a narrow channel at the surface with the characteristics of an n-type semiconductor. This is known as an inversion layer, and is shown in Figure 4.4. The presence of this layer means that electrons can flow freely from one side to the other without having to cross a p-n junction.

We can therefore see that the current flowing through the structure can be regulated by applying a small voltage to the gate electrode. In particular, a small positive voltage applied to the gate allows a current to flow freely, and we say that the device is in the ON state. If the voltage is removed there is no current flow and the device is OFF.

The terms describing the various parts of the FET have watery connotations, and so you may find the analogy with the water FET useful in order to remember

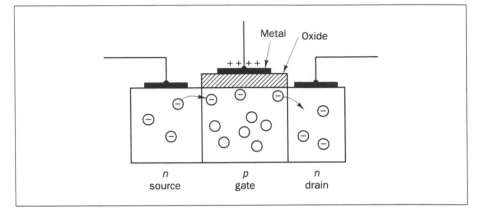

Figure 4.4 Applying a positive voltage to the gate of a MOSFET allows a current to flow through the device.

them. The negative n-type layer from which the electrons start is called the source, the intermediate p-type layer is the gate, and the positively charged n-type layer where the electrons flow out is called the drain.

The other principal type of transistor is a bipolar transistor, so called because it uses carriers of both polarities. (In comparison, the FET makes use of only one type of carrier, and could therefore be called a unipolar transistor.) We again have three layers of semiconductor, this time with a very thin layer of p-type material sandwiched between two thicker n-type layers. Unfortunately, having just been introduced to the terminology for the MOSFET, we now find that these regions have quite different names in the bipolar transistor. They are referred to as the emitter, base and collector, respectively. (In case you forget which terminology goes with which device they are labelled in Figures 4.3 and 4.5.)

Let us first consider how a bipolar transistor can be used to amplify a signal. Suppose that we apply a negative voltage to the emitter, keep the base neutral, and apply a positive voltage to the collector. As in the case of the FET, we have one forward and one reverse biased junction. The junction between the base and the collector is reverse biased. We therefore expect only a trickle of current to flow into the collector, because of the presence of minority carriers (conduction electrons) in the base. However, the junction between the emitter and the base is forward biased, so the negative voltage applied to the emitter pushes a large number of conduction electrons into the base. (This is the diffusion current that we introduced in the previous chapter.) Once in the base these electrons become minority carriers, and because the base region is very narrow, most of the electrons manage to pass right through this region without encountering a hole. What happens when they reach the other side of the base? We have just said that the applied voltage at this junction tends to favour the movement of minority

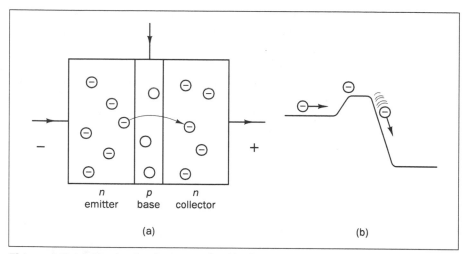

Figure 4.5 (a) The basic elements of a bipolar transistor, and (b) the effect on the conduction band energy.

carriers into the collector. Consequently, these conduction electrons are swept across the junction. We therefore find that the vast majority of the electrons moving out of the emitter succeed in reaching the collector, and so a current flows through the device.

An alternative way to describe this process is in terms of the energy picture that we introduced when describing the p-n junction. By applying a forward bias between the base and the emitter we reduce the size of the energy step at this junction, whereas the reverse bias between the base and the collector increases the energy step on the other side of the device. This is illustrated in Figure 4.5(b). We can therefore see that there is a net flow of electrons from the emitter into the base. Once the conduction electrons are in the base they rush into the collector, where there is an extremely attractive (lower) energy state.

What have we achieved by this? A current of electrons flows out of the emitter, and since only a few of these electrons recombine in the base we have a current of nearly the same magnitude flowing into the collector. There is also a tiny current of holes flowing into the base from the external contact. This is necessary because when recombination does take place, the external circuit has to restore the level of holes in this region in order to keep it electrically neutral. The key to using the device as an amplifier is to use this base current to control the collector current. To do this it is helpful to look at things from a slightly different viewpoint. Let us instead consider the effect of applying a small current of holes to the base. These excess holes introduce a small positive charge into the base. Electrons are therefore induced to flow from the emitter into the base in order to cancel this positive charge. (Remember that although the collector also has a large supply of conduction electrons, very few of these will manage to enter the base because the junction is reverse biased.) Since most of the electrons from the emitter flow right through the base without recombining with a hole, the number of electrons flowing from the emitter must greatly exceed the number of holes flowing into the base. For example, if we suppose that only one percent of the electrons recombine with a hole, then the other ninety nine percent will continue into the collector. Consequently, if we vary the base current, the current of electrons flowing from the emitter to the collector varies accordingly, in direct proportion to the current flowing in at the base. The difference is that the collector current is much greater than the base current—in this example ninety nine times larger. It follows that if we use this device in a situation where the base current serves as the input and the collector current as the output, then the transistor acts as an amplifier.

Many people struggle with the idea that a transistor can convert a weak signal into a strong signal. There seems to be something inherently wrong about this. We never get something for nothing, and the transistor is no exception to this rule. A moment of thought will show that the large current is already present. It is the current which is supplied to the emitter of the transistor. The function of the transistor is merely to imprint the pattern of the weak signal on to this large current to produce a much stronger signal. This imposes an obvious limitation on the amplifying properties of the transistor. If the base current is too large then the voltage across the emitter will not be sufficient to supply the required number of electrons. There is therefore a maximum value of the collector current for a transistor in any given circuit. Once this maximum is reached any further increase in the base current will not be reflected in the output, and we say that the device is saturated.

Following these same principles it is quite straightforward to see how the bipolar transistor acts as a switch. If a current is applied to the base, then electrons flow from the emitter to the collector and we can consider the transistor to be in an ON state. On the other hand if no current is applied to the base, then there is no reason for any electrons to enter the base, and so the device is in an OFF state.

We have now examined both of the principal kinds of transistor. We might well ask why there are two different types. Which one is the best? In fact, both have advantages in various applications, but we will confine our attention here to their use in computers. Let us first take a brief look at the history of these two technologies. The concept of a field effect transistor was first proposed and patented in the 1930s, but it was the bipolar transistor which was first to appear in working form. Developments at Bell Laboratories in the late 1940s led to the production of a point contact transistor, closely followed by the superior bipolar transistor for which John Bardeen, Walter Brattain and William Shockley received the Nobel prize. In the early 1950s Shockley resurrected the idea of the field effect transistor, but it was not until 1962 that the first working MOSFET was constructed. By this time the bipolar transistor was a mature technology and the production of integrated circuits was well under way. However, the new MOSFET proved to be more compatible with the production of planar integrated circuits: they were easier and less costly to manufacture, and could be packed more densely on to the surface of a silicon chip. (We will examine these properties in more detail in the next chapter.) These advantages have led to MOSFETs becoming the dominant technology in most integrated circuits designed for switching applications. However, bipolar transistors have the distinct advantage that they can be made to operate faster than their MOSFET counterparts. This is largely because the width of the base region in a bipolar transistor is much narrower than the gate width in a MOSFET. As we shall see shortly, this has a great influence on the maximum rate at which the device can be switched ON and OFF. Consequently, bipolar transistors still have a place in the processing units of state-of-the-art supercomputers where speed is of the essence.

Both of the transistors described here have a central region of p-type silicon. The convention for referring to these systems is that bipolar transistors are described by the ordering of the layers, in other words the type described is an npn bipolar transistor. The field effect transistor is referred to as an n-channel MOSFET, because when the device is ON the layer of material directly under the gate has the properties of n-type silicon. However, it is also possible to make the base or gate region n type and the other regions p-type to form a pnp bipolar transistor or a p-channel MOSFET. These work in the same way as the devices discussed above except that the polarities of the applied voltages are reversed, and it is holes instead of electrons which move across the base or gate. In general all of the transistors on a single chip will be of the same type, but an interesting possibility arises if we combine both p- and n-channel MOSFETS. We can see the advantage of this set-up by considering an analogous situation. Funicular railways are a common sight at seaside resorts where they are used to take people up and down a steep cliff. The arrangement consists of two carriages joined together by a cable. Initially one carriage is at the top of the incline and the other at the bottom. As one goes down it helps to pull the other one up. The carriages always operate as a pair even if one of them is empty, because it uses far less energy than operating the carriages individually. A similar idea can be used to save energy in an integrated circuit. A MOSFET uses most power at the moments when it changes state either from OFF to ON, or vice versa. Now suppose that we arrange the MOSFETs in pairs consisting of one n-channel and one p-channel MOSFET. If a negative voltage is applied to the gate of each one then the p-channel device is switched ON and the other one is OFF. If we now make the voltage positive then the p-channel device switches OFF, but at the same time the other device switches ON. In this way the pair of transistors use far less energy than if a single device is used. Such circuits are referred to as complementary or CMOS. Their low power consumption makes them especially useful in battery-powered applications such as calculators and the counting circuitry for digital watches. They are also employed in densely packed computer circuits where the need to minimize the amount of power consumed is equally important.

We have now seen how the transistor switches between two distinct states. One of the properties we will be most concerned with in later chapters is how rapidly it can perform this operation. A major factor in this respect is the average time that it takes for a carrier to cross the gate or base region of the transistor. We can obtain a feeling for this by considering the water FET. If we push down on the sluice gate it takes a certain amount of time for the water to flow from the source to the drain. Consequently, there is a time delay between the action of applying pressure to the gate and the flow of water issuing from the drain. Similarly, when the pressure on the gate is released, some of the water which is in transit will still reach the drain, and so the flow of water will continue for a short time. This delay in the response of the system is of no concern so long as the periods of time during which the gate is depressed or released are significantly longer than the delay time. However, we run into problems if we try to operate the system faster than this. For instance, let us suppose that the gate is pushed down, but then released again quickly before the water has had time to reach the drain. Some of the water may dribble into the drain, but is it sufficient for us to recognize that the gate has been opened? We get a similar problem if we release the gate and then depress it again very shortly afterwards. By use of a similar argument we find that the shortest pulses which can be used to switch the transistor ON must be long enough to allow a number of electrons (or holes) to pass through the channel under the gate before the channel is destroyed. A great deal of effort is being made to reduce this time period, and we will return to this in the following two chapters.

*

For the remainder of this chapter we will briefly examine how transistors are used to perform the three essential functions required for a computer, namely the ability to evaluate logical and arithmetic functions and to store information. We will not go into great detail as this would take us too far from the central theme of the book. I have tried to incorporate just the minimum discussion on this subject which I think necessary to appreciate the workings of a computer, but those readers who find the following too involved or tedious may jump to the end of the chapter without significantly affecting their understanding of the rest of the book. We will consider examples employing MOSFETs which are generally easier to understand, but similar circuits can also be constructed using bipolar transistors.

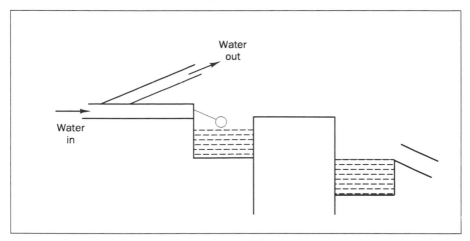

Figure 4.6 A NOT gate based on the water FET. It is shown with the transistor in the OFF state.

The rules of logic employed by all electronic computers are based on those devised by the nineteenth-century English mathematician George Boole. The system requires each logical statement to be either TRUE or FALSE. This immediately suggests a link with transistors which also have two distinct output states, and so we can arbitrarily associate the logic levels TRUE and FALSE with the transistor being in the states ON and OFF, respectively. In fact, this link was first made by Claude Shannon in 1938 for use with electro-mechanical switches, before the advent of the semiconductor transistor. We will begin by looking at the simplest logic function, NOT, the output of which is simply the opposite of the input. We can see how this function is performed by examining the water FET once again. Consider the arrangement as shown in Figure 4.6. It is similar to the previous diagram of the water FET (Figure 4.2) except for the presence of an additional pipe which is taken off horizontally from the supply pipe. If the water FET is switched OFF, meaning that there is no pressure on the sluice gate, then the water cannot enter the top tank. Instead it flows out through the horizontal pipe, which we shall call the output pipe. On the other hand, if pressure is applied to the sluice gate, then water is allowed to flow into the top tank, over the gate into the lower tank, and out through the overflow, taking this route in preference to the output pipe. So whether or not an input is applied to the sluice gate determines whether the water flows through the output pipe or goes to waste down the drain. We can interpret these results as follows. If there is no pressure on the sluice gate, in other words the input is FALSE, then water flows through the output, so the result is TRUE. Similarly, if the input is TRUE, then the absence of water in the output pipe indicates a result of FALSE. This is just what we required of a NOT function. A MOSFET can be used in

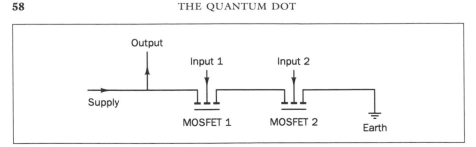

Figure 4.7 A NAND gate formed by placing two MOSFETs in series.

a similar arrangement, except that we have a flow of electrons, not water. In this case electrons flow through the output if no voltage is applied to the gate, but travel out through the drain to earth if the device is switched ON.

A device, or sequence of devices, which performs a logic function is called a logic gate. (This is slightly confusing terminology as we also have a gate region of a FET which is something entirely different, but from the context it should always be quite clear which meaning is intended.) The NOT gate requires a single input and just one transistor, but most other logic gates require two inputs and several transistors. Two common examples are the AND and OR gates. The function of these is self-explanatory. The AND gate returns a TRUE output only if both of the inputs are TRUE, whereas an OR gate requires that either one or both of the inputs are TRUE. These three gates, AND, OR and NOT, are sufficient for evaluating any logical expression, but in practice many other variants are used. An example is the NAND gate. This is an amalgam of NOT and AND, and so produces a TRUE output in all cases except when both inputs are TRUE. Figure 4.7 shows how such a gate can be constructed using two MOSFETs in series, one after the other. One of the inputs is connected to each of the transistors, but the current can only flow through the transistors if they are both in the ON state. Otherwise it will simply flow through the output wire. We therefore achieve a response which satisfies the criteria for the NAND function. (If desired, an AND gate could easily be achieved by using the output of this device as the input to a NOT gate.)

Let us now look at how transistors can be used to evaluate arithmetic expressions. Firstly, we need to understand how numbers are represented in a computer. As we discussed at the start of this chapter, computers perform numerical calculations in binary. It is quite straightforward, although somewhat tedious, to transfer numbers in base ten into those in base two, and vice versa. A number in base ten is made up of a units column, a tens column, a hundreds column, and so on. For example, the number 364 represents 4 units, 6 tens and 3 hundreds. Binary numbers are constructed in a similar way except that the columns from right to left represent increasing powers of two. So, for example, the binary number 1101 has one unit, nothing in the twos column, and a one in both the fours and the eights column, making a total of thirteen in base ten.

This is all right for dealing with whole numbers, or integers, but most calculations involve real (i.e. non-integer) numbers. How do we represent a number such as 89.1 so that it can be understood by a computer? The most convenient way to do this is to use the so-called floating point representation which is similar to that used in the display of scientific calculators. In this system each number is represented by two integers, called the mantissa and the exponent. The mantissa is identical to the original number with the decimal point removed. So, for example, the mantissa associated with the number 89.1 is 891, which the computer interprets as 0.891. The other part of the number, the exponent, gives the power of ten by which the mantissa must be multiplied to regain the original number. Or, more simply, we can define the exponent as the number of digits on the left-hand side of the decimal point in the original number. Consequently, in this example the exponent is 2. We can easily see that this is consistent with the former definition since if we multiply 0.891 by 10 to the power 2, or 100, we regain the number 89.1.

One feature of our own number system which we often take for granted is that it is relatively compact. This is not usually appreciated until we consider how many digits are required to represent a number in binary. For example, the number 891 in the above example requires ten binary digits or 'bits'. On top of this we also need other bits to store the exponent and one more bit to tell us whether the number is negative or positive. (We will return to the subject of negative numbers shortly.) Consequently, a large number of bits are required to store a number with sufficient accuracy. Typically, to retain the six most significant decimal figures in a number it is necessary to use thirty-two bits. In fact, most supercomputers use sixty-four bits to achieve an accuracy of fourteen decimal places. This may seem excessive, but when many repetitive operations are performed the errors in the less significant digits can accumulate alarmingly so that the final answer is far less accurate than we might expect.

Having considered how we can represent numbers on a computer, we now want to be able to manipulate them. One of the simplest tasks is to add two numbers together. Just as in base ten, we can add two binary numbers together by adding each column, starting with the least significant, rightmost digits. The rules for binary addition are very simple when we have just two digits. If both are 0 then the answer is also 0; if one is 0 and the other is 1 then the answer is 1. The only complication arises if both of the digits are 1. The result is 2, but in binary this is written as 10, so we would write a result of 0 and carry the 1 into the next column. If we ignore this carried digit for the moment, then there is a very limited set of possibilities. There are two inputs, each of which is either 0 or 1, and two possible values of the output which are also 0 or 1. We could achieve this same result with an appropriate logic gate if we associate the states

Figure 4.8 A logic gate circuit for adding two inputs.

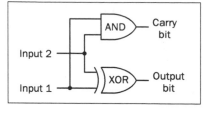

0 and 1 with FALSE and TRUE, respectively. The relevant gate is called an exclusive-OR, or XOR, gate. We can think of this as behaving in a similar way to a two-way light switch. The bulb lights if one of the switches is turned on, but will go off if the other switch is also turned on. Similarly, the XOR gate produces an output of TRUE if one switch is on and the other is off, but will be FALSE if both are off or both are on. To construct such a logic gate requires several MOSFETs, but we will not worry about the intricacies of how this is achieved. Next we need to consider how to generate the carry bit. This is straightforward since we know that the only situation in which this value is 1 is if both of the inputs are 1, a condition which is achieved with an AND gate. The resulting structure is shown in Figure 4.8.

Such a circuit is only sufficient for dealing with the rightmost column of digits where there are only two digits to add together. If we look at the next column then there is also the carry bit from the previous column to consider, giving a total of three digits. However, the sum can still be performed using logic gates with only two inputs simply by adding two of the digits together first and then adding the third digit onto this sum.

The process of performing addition in binary is therefore a specialized case of using logic gates. However, a computer must perform far more than simple addition. As a minimum requirement it must also be able to multiply, subtract and divide two numbers. Fortunately, in binary these processes are closely related to addition. Let us look first at multiplication. For example, suppose we wish to multiply together the two binary numbers 1101 and 11. The right-hand digit of the multiplier, 11, is a 1, so when we multiply this by the multiplicand, 1101, we simply get the number 1101. We store this in a register called the accumulator. Next we need to multiply by 10, the result of which is similar to multiplying by the usual number ten: it simply shifts the multiplicand one space to the left, so we obtain 11010. This is then added to the number in the accumulator. In this case the accumulator now contains the desired result. However, much larger numbers can also be handled simply and efficiently using the same process of repeated addition.

To understand how subtraction is performed we need to take a closer look at the way in which negative numbers are represented. We have said that one bit of the binary number is used as a sign bit; this is the leftmost bit, and the convention is that a 0 indicates a positive number and a 1 a negative number. However, the most convenient form for performing subtraction is to represent a negative number by what is called the twos complement. This is obtained by changing all of the digits (including the sign bit) so that 1s become 0s and vice versa, and then adding 1. For example, the four-bit representation of +3 is 0011 where the leftmost 0 indicates that the number is positive. To obtain the number −3 we switch all of the 1s and 0s to give 1100 and then add 1, giving 1101. As expected, the leftmost digit is a 1, indicating that the number is negative. However, the interesting feature of a twos complement number is that it allows us to subtract one number from another by performing addition. This is obvious when you think about it: in base ten the sum 6 − 3 is of course equivalent to writing 6 + (−3). In binary the positive value 6 is written as 0110 and the number −3 as 1101. If we add these we get 10011, but the leftmost bit will overflow the register and so we are left with 0011, or +3, as expected. Finally, division can be performed by repeated subtraction (although in practice there are much faster methods).

From the above discussion we can see that multiplication, division and subtraction can all be performed by one piece of hardware which simply adds two numbers together. In practice, it is far more efficient to construct additional units which are dedicated to performing multiplication and division. All of this requires a large number of transistors. For example, a unit which adds together two binary digits (plus a carry bit) consists of about twenty transistors. If we are dealing with sixty-four bit numbers then this arrangement needs to be replicated sixty-four times. Once we include units that multiply and divide, as well as ones that perform logical functions, the number of transistors required runs into many thousands. However, this number is tiny in comparison to the number needed to form the memory of the machine.

Computer memory comes in several forms. It is used to store input and output data as well as holding intermediate values of calculations. We have already seen an example of this in the accumulator register. The memory also contains the program itself. In fact the great versatility of the computer comes from the fact that the instructions which determine how the computer processes the data are held in the memory, just like any other data. For users of microcomputers the most familiar form of memory is the floppy disk. However, anyone who has ever sat at a computer waiting for a program to be loaded from a floppy disk will appreciate that it takes the computer a considerable amount of time (at least a few seconds) to access the data. Since we expect even a microcomputer to perform calculations at a rate approaching one every millionth of a second (and supercomputers to be several hundred times faster than this), it is necessary to have much faster access to the data. The solution is to construct memory cells from the transistors themselves. Some of these cells, such as those incorporated in the accumulator, must be included on the same chip as the arithmetic unit,

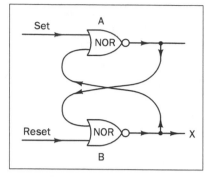

Figure 4.9 A flip-flop circuit which can be used as a memory element.

while others should be on neighbouring chips. In fact, many supercomputers possess a hierarchy of memories, the fastest ones being located as close as possible to the arithmetic unit, while those placed in more distant locations have longer access times.

The problem of using transistors to store data is that they have no sense of history. We have seen for example that if we apply a voltage to the gate of a MOSFET it switches on, but when the voltage is removed it switches off agaii It retains no record of the fact that it has ever been switched on. What we neec to be able to do is to store or 'latch' a signal. The basis for doing this is a simple circuit called a flip-flop. This has nothing to do with the plastic shoes you wear on holiday. In its most basic form it is simply a loop containing two logic gates. The output of one gate forms one of the inputs to the other, and the output from this one forms one of the inputs to the first.

Consider the arrangement shown in Figure 4.9 consisting of two NOR gates which we have labelled as gates A and B. One of the inputs to gate A is the output from gate B, and the other we will call the Set input. Similarly the output of gate A forms one of the inputs to gate B, and the other we will refer to as Reset. To follow this discussion it is simply necessary to remember that the output of a NOR gate is 1 only when both of the inputs are 0. Suppose that initially the Reset signal is 0 and a short voltage pulse is applied which produces a 1 at the Set input. Since this input to gate A is not 0, it follows that the output from this gate must be 0. This is one of the inputs to gate B, the other being the Reset signal, which is also 0. Consequently, the output from gate B, and that which we would observe in the output of the flip-flop at X, is a 1. Since we now have a 1 as input to gate A, the output from A will be 0 regardless of the input on the Set line. This means that when the Set signal returns to zero the state of the system will remain unchanged. We will continue to have an output of 0 from gate A and a 1 from gate B. The system has therefore 'remembered' that a signal was applied briefly to the Set channel. This memory state persists until a Reset signal is sent. When a voltage pulse is applied to the Reset channel there is then an input of 1 at gate B. This means that the output from gate B must be 0.

Assuming that the Set signal is 0, then both inputs to gate A are 0 and so the output is 1. Once again we find that this new state persists after the Reset signal is removed. This outlines the general operation of a memory cell. In practice many refinements are made, but the basic principles remain the same.

We have now reached a point where we understand the basic principles governing the operation of the semiconductor transistor. Starting from the ideas of discrete electron energy levels in atoms and the formation of allowed energy bands in solids, we have explored the unique properties of semiconductors and seen how these can be altered by introducing small quantities of foreign atoms. By forming a junction between two regions of semiconductor with different types of doping we have seen that a device with asymmetric characteristics is produced. From this concept we have arrived at a structure in which the flow of electrons through the device can be dramatically altered by applying a small voltage or current to one of the terminals. This allows it to be used either as an amplifier or as a switch. It is the switching behaviour, the ability to be either ON or OFF, which we will concentrate on throughout this book, but to close this chapter we will briefly compare the two types of behaviour.

As an amplifier the transistor acts as an analogue device, while as a switch it forms the basis for all digital electronics. An example of a transistor as an analogue device is reflected in the behaviour of a typical hi-fi amplifier. The input and output of such a system vary continuously. This means that the tone and volume may change by arbitrarily small amounts. The aim of the amplifier is to magnify the signal as faithfully as possible in order to maintain the relative magnitudes of all the different harmonics. However, even the best amplifier introduces some distortion into the signal. Worse still, it is subject to 'noise', which means that any unwanted fluctuations in the input signal are also amplified. In contrast, digital circuits are particularly immune to noise since it would require a spurious signal large enough to change an OFF signal into an ON signal in order to produce an error. Although audio and video signals are conventionally thought of as being analogue, the advantages of digital systems can also be applied in these contexts. Once digitized, these signals can be treated as numbers, and so can be processed in the same way as any other digital information on a computer. However, the vast quantities of data involved present serious problems. For an audio signal the problem is not too severe, although it is necessary to represent the volume of the sound at a large range of wavelengths for successive short intervals of time. The problem is much worse when dealing with optical images where it is necessary to break the image down into small pixels and store the intensity and colour of each pixel for every time step. This is one of the most challenging problems facing the world of digital electronics, and we will return to it later in the book.

The Amazing Shrinking Transistor
The Benefits of Integrated Circuits

THE introduction of the semiconductor transistor revolutionized the computer industry and had far-reaching effects in many other areas. However, the massive reduction in both the size and cost of electronic devices was due to a further development, that of the integrated circuit. Without this, many of the applications which we now take for granted —pocket-sized calculators, personal computers, and the appearance of electronics in everything from washing machines to children's toys—would not be possible. The scale of this advance is hard to imagine. Just over thirty years ago electrical circuits were composed of individual discrete devices, each measuring typically several millimetres across. Nowadays it is possible to place several million devices on to a thin slice of silicon about the size of a thumbnail. The sizes of the individual components are measured in microns, a unit of length equal to one thousandth of a millimetre. How is it possible to make such tiny circuits? And how much smaller can they be made before we encounter fundamental limitations? These are two of the questions we will address in this chapter, but first of all we need to ask a more basic question: what is an integrated circuit?

We will answer this by first considering an ordinary electrical circuit, for example the kind we would expect to find if we opened up a thirty-year-old hi-fi amplifier. Typically there are a large number of resistors, a selection of capacitors, inductors and diodes, and a handful of bipolar transistors. The whole collection is mounted on a circuit board and the individual components are joined by wires, each one being attached to the board by a small blob of solder. In contrast, if we prised open the packaging around an integrated circuit we would find something quite different. It would be much smaller in size than the circuit board, so we may need a magnifying glass, or even a microscope, to identify the individual components, but even then it would be very hard for the untrained eye to recognize any particular device. This is because they are not simply miniature versions of their discrete counterparts. The individual elements of the integrated circuit are effectively confined to the two-dimensional plane that is

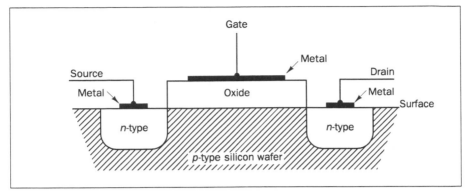

Figure 5.1 A planar MOSFET as implemented in an integrated circuit.

the surface of the chip, and are interconnected by thin metal lines deposited on this surface. Thus the whole structure is flat, or planar. How does this affect the components?

One restriction is that inductors, which are normally made from a cylindrical coil of wire, can not be fabricated in a form suitable for inclusion in an integrated circuit. This means that we can not take a design for a circuit consisting of discrete components and simply reproduce it in an integrated form. However, this is the last thing we would want to do as the integrated and discrete circuits are designed to fulfil very different tasks. As we discussed at the end of the previous chapter, the discrete amplifier circuit is an analogue system, and many of the inductors and capacitors are used in connection with minimizing the distortion produced by amplification. In a digital system these parts of the circuit are not required, and so we would in any case expect to have far fewer inductors and capacitors. Fortunately, it has proved possible to eliminate the need for inductors altogether. With a bit of ingenuity capacitors and resistors can be constructed in planar form, as can bipolar transistors. However, all of these tend to require quite large amounts of surface area. The device most suited for use in an integrated circuit is the MOSFET. This relies on surface effects for its operation and, as we can see from Figure 5.1, can easily be constructed so that all the electrical contacts are on the top surface. It is also very efficient in terms of the amount of surface area of the chip that it occupies. This is of principal importance since it directly affects the number of devices which can be incorporated on to a single chip. Indeed, since a MOSFET usually requires less space than a resistor in an integrated circuit, it is quite common for resistors simply to be replaced by MOSFETs.

We therefore find that integrated circuits are very different from circuits constructed from discrete components. In particular, transistors are used only where necessary in discrete circuits, as they are more expensive to produce than resistors and other components, whereas they tend to be the most prevalent type of device in integrated circuits where the cost of a device is measured in terms of the surface area that it occupies.

✶

How do we go about making an integrated circuit? The starting-point is to design the circuit. This can be a very lengthy process. Producing a plan for an integrated circuit containing upwards of a million transistors is a task akin to planning the layout of a medium-sized city down to the detail of each individual house starting from scratch. However, there are now many computer packages which assist in this process. Once the design is complete the actual production can begin. The first stage is to produce the wafer, a thin disk of silicon, typically less than half a millimetre thick and about 20 centimetres in diameter. This is made from exceptionally pure silicon to which a single type of doping, say p-type, has been added.

The next stage is to convert the plan of the integrated circuit into a real circuit on the surface of the wafer. The technique is similar to one we could use to transfer a complex colourful pattern on to a blank sheet of paper. We first identify all of the red areas in the pattern and then make a stencil with windows corresponding to these areas. If we lay the stencil over the blank sheet of paper we can then simply spray the whole area with red paint. The process can be repeated for the different colours until we have reproduced the entire picture. The beauty of this method is that once we have made the stencils, the pattern can be generated over and over again with very little effort. In a similar way the plan of the integrated circuit must be first broken down into a series of stages, each one requiring a particular type of processing. Let us take a detailed look at one such stage of the fabrication process. For example, we might first identify all those areas which need to be made n-type. A photographic image of this master plan is made and then reduced to the desired size. (We can consider this process to be the opposite of that normally used to enlarge a photographic negative.) Since this image now measures typically only a few millimetres along each side, it covers a very small portion of the wafer. The pattern can therefore be repeated until it covers the entire surface of the wafer. The finished article is called a mask and is an image consisting of several hundred identical circuits side by side. In the following stages all of these circuits are fabricated simultaneously—a striking example of mass production.

We next have to treat the surface of the silicon wafer. It is first oxidized by placing it in a furnace in an oxygen-rich atmosphere. This creates a thin layer of silicon dioxide on the surface, the same material which forms the insulator layer in the metal-oxide-semiconductor transistor. On top of this is placed a thin coating of resist, a material which is sensitive to ultraviolet light. This acts like a photographic film. The mask is placed on the surface of the wafer and illuminated with ultraviolet light. When the 'photograph' is developed, the unexposed resist is washed away leaving an image of the mask on the surface. The final stage of this patterning process is selectively to etch away these regions of the oxide to reveal the silicon surface below. This is achieved by immersing the wafer in an acid which attacks the oxide but not the hardened resist. In this way the unprotected oxide is eaten away and the newly exposed silicon surfaces correspond to the areas defined on the mask. This process of transferring the pattern is called photolithography. It is repeated several times in the manufacture of the integrated circuit, usually once for each processing step.

Having defined the pattern on the surface of the wafer, we now treat it accordingly. For example, to make these regions n-type we must introduce some donor atoms. Since we assumed that the wafer initially contained acceptor doping, some of the conduction electrons introduced will recombine with the holes which are already present, a process known as compensation. We must therefore introduce a sufficient quantity of donors to produce a majority of conduction electrons. There are two principal techniques. One is to place the wafer in a furnace containing a gas of suitable donor atoms. At a temperature of about a thousand Celsius these donor atoms are able to diffuse slowly into the exposed silicon. An alternative method is ion implantation, in which the donors are accelerated to high speeds and fired into the wafer. In this case, the oxide layer must be thick enough to ensure that the ions cannot penetrate through the oxide into the silicon below.

Several other processing stages may be involved, with a new oxide layer and resist being grown in each case. The final step is to open up small windows in the oxide in order to make the electrical contacts. These are formed by depositing a thin film of metal, usually aluminium. The insulating properties of the oxide layer are again important in this case, ensuring that contact is made only at those points where the silicon is exposed.

The finished wafer consists of a few hundred identical circuits, each of which will typically contain several hundred thousand components. However, not all of these circuits will work. Tiny defects in the silicon wafer may cause one or more of the devices to malfunction. Cleanliness is also of enormous importance in the manufacture of these structures. Extraordinary measures are taken to provide clean conditions, but even then small dust particles a few microns across may contaminate a circuit. Since the size of these dust particles is comparable with the width of the contact lines used to connect the devices, a single such particle may interrupt one of these connections and cause the whole circuit to fail. Each circuit is therefore tested by an automated process. Those which pass are packaged ready for use, while those which fail are simply thrown away—it is not economically feasible to repair them.

As we have already stated, the MOSFET is particularly well suited to the planar

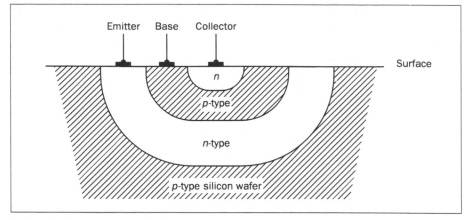

Figure 5.2 A planar bipolar transistor.

approach required for an integrated circuit element. It requires only one stage of doping to create two small regions of n-type material which act as the source and drain, and by making an electrical contact to the surface of the oxide above the intervening region we define the gate region of the device. In contrast, bipolar transistors require a greater degree of complexity. As we shall see in the following discussion, this is chiefly because they rely on the transit of minority carriers.

We could of course create a single bipolar transistor by simply forming two small regions of n-type material as for the MOSFET. The problem comes when we try to create other bipolar transistors in close proximity to this one. There is then nothing to stop the electrons from the emitter of one transistor finding their way into the collector of another one. In order to avoid this we need somehow to isolate each bipolar transistor. The usual way of achieving this requires two or more processing steps. We could begin by doping a large, comparatively deep region of the wafer with donors, and then introduce two small regions of p-type doping within this region to form the emitter and collector. (Of course, this forms a pnp transistor. If we wish to form an npn transistor as before we must start out with an n-type wafer and use the opposite types of doping at each stage.) However, this is still not a particularly satisfactory arrangement since the base region of a bipolar transistor must be very narrow, requiring that the small p-type regions be placed very close together. A better way to meet this criterion is shown in Figure 5.2. In this case three processing stages are required to produce successively smaller islands of alternately doped material. In this way the outer n-type region acts as the emitter, the p-type layer is the base and the small n-type region in the middle is the collector. This is referred to as a 'vertical' device since the current flows in a direction perpendicular to the surface of the wafer. It is quite clear that in this case an electron from the base of one transistor cannot reach the collector of a different one. However, this structure uses up more surface area than a comparable MOSFET and requires a far more complex manufacturing process. The combination of these features makes the bipolar integrated circuit considerably more expensive to produce than a similar MOSFET circuit.

There are several advantages to integration, some of which we examined briefly at the beginning of this book. One feature is the improvement in reliability since the electrical contacts within an integrated circuit are far less likely to fail than the conventional soldered joints used to attach discrete components on to a circuit board. Assembling components to form a circuit is also an expensive process, especially when one considers the testing and resoldering of defective joints. This desire to minimize the cost of an electronic circuit, usually expressed as the cost of performing an electronic function, has been the main driving force behind the ever-increasing levels of integration. For example, the cost of a memory element constructed from discrete devices would be equal to the total price of the individual components plus the cost of soldering them on to a circuit board. In contrast, an integrated circuit might contain a hundred thousand memory elements, in which case the cost of a single function would be one hundred thousandth of the cost of producing the circuit. It seems then that the way to reduce the cost per function is to increase the number of devices on a single chip, although there are certain pitfalls as we shall see.

The goal of increasing the number of devices in a single circuit has been tackled from two directions. The simplest is to allow the area occupied by each circuit, a region known as the die area, to increase. The main problem encountered with this approach is obtaining a high yield, in other words ensuring that a large proportion of the finished circuits work. Consider an example in which four hundred circuits are placed on a wafer. We discover that due to various defects one hundred of these do not work, so there is a seventy-five percent success rate. If the cost of processing a wafer is £3000, then each circuit costs £10 to produce. Suppose that we now double the area of the die. It is now possible to fit only two hundred circuits on the wafer. (In fact, the actual number may be slightly less than this since the dies are square and the wafer is circular, so some will be lost at the edge.) If there are still one hundred defects then the success rate will be only fifty percent. (Again we could argue that there may be slightly less than 100 defective circuits, since some of these larger circuits may incorporate two defects, but this will be compensated by the fact that there are less than two hundred complete circuits.) The cost per circuit is now £30, thus although the larger circuits may contain twice as many devices, the cost per function is actually more than for the smaller circuit. This argument seems to suggest that the lowest cost per function is obtained with very small circuits, contrary to what we have stated. However, when the packaging costs of, say, £30 per circuit, are taken into account, the total costs of the circuits become £40 and £60, respectively. It is then found that the larger circuit does indeed give a lower cost per function, but there are clearly limitations on how far this approach remains viable. It does not take a great deal of thought to see that if the die area is doubled again there will at best be only a handful of working circuits, the cost of which will be prohibitively expensive. There is, therefore, a trade-off between maximizing the yield of good circuits and at the same time placing the most devices of a given size into a single circuit. Nevertheless, the die area has increased steadily over the years with improvements in the quality of the silicon wafer and in the clean room facilities. As a result the area of a circuit has increased by approximately one hundredfold over the past thirty years without significantly affecting the yield.

At the same time as the circuit size has been increasing, the size of the individual components has decreased substantially. This has been achieved primarily by progressively refining the photolithographic techniques, allowing finer structures to be resolved. The benefits of this reduction in size are easy to appreciate. If we reduce all of the dimensions by a factor of two, then four times as many devices can be squeezed on to the same area as before. Since the cost of a circuit is effectively governed by the surface area that it occupies, this reduces the cost per function by a factor of four. There is also a second major incentive to reducing the size of the individual components, and that is the effect it has on the speed of operation of the devices. As we saw in the previous chapter, the switching speed of a MOSFET is governed by the time it takes an electron to cross the gate region. By reducing this distance by a factor of two we therefore also obtain a similar increase in the performance of the devices.

These advantages of reduced cost, smaller size and increased performance have fuelled the rapid increase in the number of electronic devices which can be placed on a single silicon chip. This is clearly illustrated in Figure 5.3, which shows how the maximum number of devices on a chip has increased since the invention of the integrated circuit in 1959. The first circuits, with typically less than fifty devices per circuit, were referred to as small-scale integration. Since then we have passed through medium- and large-scale integration and have now arrived at the state-of-the-art technology called simply very-large-scale integration, or VLSI. Although there are no hard and fast boundaries, such a circuit is generally defined as containing upwards of one hundred thousand transistors. How much further can we go? Common sense suggests that there must be some limit, particularly with regard to reducing the sizes of the individual devices. Nevertheless, circuits with several million components have already been manufactured, taking the technology into the realms of ultra-large-scale integration. (Thankfully, this terminology seems to have usurped the unimaginative title very very large-scale integration!) Circuit designers are now talking about giga-scale integration, representing one billion devices on a chip, shortly after the next turn

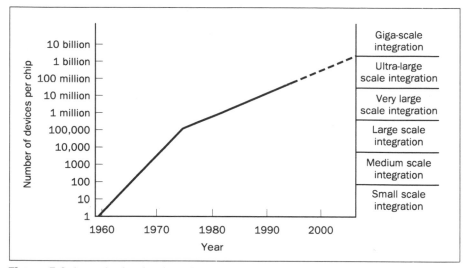

Figure 5.3 A graph showing how the number of devices on an integrated circuit has increased since 1959, and the predicted trend to 2005.

of the century. How will this be achieved? Let us consider this question by looking at the difficulties associated with further reducing the sizes of the individual devices.

First of all we should maybe consider whether it is possible to manufacture devices of the desired size. The main limitation in this context comes from photolithography—the process of transferring the circuit design on to the surface of the chip. In the past, successive refinements in the photolithography process have enabled it to keep pace with the demands imposed by increasingly smaller components. The minimum feature size which can currently be achieved in a commercial situation is about half a micron, whereas giga-scale integration will require MOSFETs with gate lengths of about a fifth of a micron. Will optical lithography, using ultraviolet light, be capable of meeting these demands in the future? The answer is uncertain, but in any event there are several other techniques which are appropriate.

One such alternative is the same as that used to produce high-powered microscopes. We dispense with the light beam and instead use a fine beam of electrons. In this process the existence of the photographic image of the circuit is not required. Instead the computer uses its stored plan of the circuit to guide the electron beam over the surface of the wafer. In this way the pattern is transferred directly on to the wafer, as though the computer is writing on the surface with the beam of electrons. The disadvantage of this technique is that the electron beam has to be steered across the surface, exposing each of the desired areas individually. This is much slower, and consequently more expensive, than using a mask which allows all of the areas on the wafer to be exposed at once. To return to the analogy we used previously, the conventional photolithographic technique is like using a spray can and a stencil, whereas the electron beam method is akin to painstakingly transferring a complex pattern using a fine paintbrush. Another possibility which avoids this problem is to retain the use of the mask, but to use X-rays instead of ultraviolet light. However, although commercial X-ray lithography systems have been developed, there are problems with the masks, which tend to be fragile, and with accurate positioning of the masks in relation to the wafer.

It is interesting to note that over twenty years ago people were forecasting that conventional photolithography would be unable to meet the future require-ments, and yet it is still the dominant technology— the alternatives that we have mentioned above have gained only a small foothold in the market. This is a phenomenon which we will meet again in different contexts: although there may be many new and potentially advantageous solutions to a given problem, the one which tends to gain acceptance in the industrial world is that which is a development of the existing technology. Consequently, in twenty years from now we may well find that photolithography is still the dominant process for transferring integrated circuit designs on to the wafer, and it may be capable of reproducing details far smaller than we currently believe possible.

Having the technology to transfer the image to the wafer is not the only problem associated with manufacturing smaller devices. As the size of the individual components is reduced, the chance of a microscopic dust particle causing a catastrophic failure of one of the elements increases. Also, since several lithographic stages are required, it is vital that each mask should be precisely aligned on the wafer. Current technology is more than able to cope with these requirements, and therefore on the manufacturing side there seems little reason why the progress towards ever smaller devices should not continue at least into the next century.

A second point concerns the devices themselves. How far can the dimensions of a transistor be reduced before it ceases to function in the desired way? We will concentrate on MOSFET structures, because they are best suited to implementation in integrated circuits, and consequently the highest levels of integration are generally achieved with these devices. The rules governing the behaviour of these structures as the dimensions are reduced, or scaled, are in principle quite straightforward—all of the dimensions, both lateral (i.e. along the surface of the chip) and vertical (i.e. perpendicular to the surface), are reduced by a common scaling factor. In the following we will consider the implications of applying a scaling factor of two. As we have already seen, this leads to a fourfold increase in the number of devices that can be accommodated within a given area. What about the other parameters?

Let us suppose for a moment that the current and voltages are maintained at the same levels as before. The electrical power consumed by each device is given by the product of the current and the switching voltage (that is, the change in voltage applied to the gate electrode in order to cause it to switch from one state to the other). Consequently, the power required by each device remains constant. However, since there are now four times as many devices within a given area, the power requirements per unit area are increased by a factor of four. This poses a serious problem since the power is converted into heat. Consequently, the amount of heat produced also increases by a similar factor. This excess heat must be dissipated somehow or the temperature of the chip will increase until something catastrophic occurs. As it is, a standard chip produces about the same amount of power as a dim light bulb. This may not sound like a terrific amount, but it is actually several times more than the amount of heat produced by the same area of an element on an electric cooker.

The usual way to combat this problem is to apply a technique known as constant field scaling. Let us take a moment to see what this means. If we think in terms of an applied voltage tilting the energy bands in the crystal (as we introduced in Chapter 2), then the electric field strength is simply a measure of the gradient of this slope. Since the distance between the source and drain has been reduced by the scaling factor of two, we need to reduce the voltage difference by a similar factor in order to obtain the same gradient as before. There are several other parameters which are also affected by constant field scaling. The switching voltage is reduced by the scaling factor, as is the current flowing through the device. Consequently, the power produced by each device decreases by a factor of four. This is ideal. It means that the amount of heat produced per unit area remains constant as the device size is scaled.

This all sounds very promising. The problems occur if for some reason one parameter cannot be scaled as desired. For instance, it is often undesirable to reduce the switching voltage because this increases the susceptibility of the system to 'noise'. What we mean by this is that even when the device is supposed to be off, a small fluctuation in the voltage may cause the device to switch on momentarily. This is most undesirable, particularly in a memory circuit where such a change of state could lead to the contents of the memory element being altered. Keeping the voltage constant, or at least reducing it by a factor of somewhat less than two, can alleviate this problem, but it returns us to the problem of heat dissipation. Another problem concerns the depletion layers associated with each of the p-n junctions in the device. In equilibrium these layers are areas in which there are virtually no conduction electrons or holes. In a large device they extend only a short distance into the gate region. In fact, when we analysed the workings of the MOSFET in the previous chapter we neglected to consider the depletion layers, assuming them to be too small to worry about. However, as the devices decrease in size, the width of the depletion layers stays approximately the same. This means that these layers occupy increasingly more of the gate region. If we reach a situation where the depletion zones from the source and drain stretch right across the gate region, then the transistor will no longer work as we expect it to. To avoid this problem is is necessary to increase the concentration of dopant atoms in the gate region. In this way there are more holes for the electrons to recombine with, and so the depletion layers do not spread as far under the gate. Since the supply voltage, doping concentration and the widths of the depletion layers are closely related, this introduces further problems. However, considering all these many criteria, most estimates of the minimum gate length in a MOSFET are around about a fifth of a micron, just what is required for giga-scale integration.

Strangely enough, the greatest problems facing further miniaturization of integrated circuits concern the humble interconnect, the tiny connecting lines which form the electrical links between the devices on the chip. Various materials are used to form these conducting elements, ranging from heavily doped silicon, to silicides and metals, principally aluminium. One of the key problems with interconnects is their resistance to the flow of an electrical current. We will concentrate on aluminium, which has the lowest resistance of the commonly used materials, since problems associated with doped silicon and silicide interconnects can often be avoided simply by replacing the material with aluminium.

In order to understand the problems posed by interconnects, and how they can be circumvented, we need to understand something about the way in which a signal travels along a wire. A popular misconception is that it requires an electron to travel from one end of the wire to the other in order to transmit a signal. However, as we shall see in the next chapter, the average velocity of an electron in the direction of current flow is very small, of the order of a few metres per second. If we had to rely on an electron travelling along the length of a wire in order to transmit a signal, then the delay in a transatlantic phone call would be several weeks! In fact, an electrical signal in a wire behaves more like a wave travelling along a stretched string. Let us consider the analogical cases first with the wave on the stretched string. We intuitively expect that a wave will travel faster along a taut stretched string than along a slack one. Similarly, we also expect the signal to travel faster along a thin string than along a thick rope. From these considerations we can show that the speed at which a wave propagates depends primarily on two factors: the tension of the string and the mass per unit length of the string. The time taken for a wave to travel along a given string therefore depends on these two factors and on the length of the piece of string. In the case of the electrical signal there are also two key parameters. These are the resistance of the wire to the flow of current, and the capacitance, or amount of charge stored in the system. The time taken for a signal to propagate along a wire is therefore dependent on what is called the *RC* factor, the product of the resistance and the capacitance of the wire. It may seem surprising that the time does not explicitly depend on the length of the wire, but there is a strong implicit dependence since both resistance and capacitance vary with the length of the wire.

The most obvious approach when reducing the size of components in integrated circuits is to scale the interconnects in the same manner, so that the width and thickness of the connecting wires are decreased accordingly. We can predict the effects of these measures by considering how the resistance and capacitance change under such scaling. Let us examine the capacitive effects first. The interconnect runs across the top surface of the oxide layer and so attracts an equal amount of opposite charges in the surface of the semiconductor below. The amount of charge stored in the interconnect depends on the area which is in contact with the surface of the oxide, and also depends inversely on the thickness of the oxide layer separating the interconnect from the semiconductor wafer. We have proposed that the width of the interconnect is reduced by the scaling factor of two, but how does the length of the interconnect change? For the time being let us suppose that it is also subject to the same factor, an assumption which should be more or less true at least for interconnections between neighbouring devices. From these considerations we estimate that the area of the interconnect in contact with the surface of the oxide is a factor of four times smaller in the scaled circuit. However, this is partially offset by the fact that the oxide layer is reduced in thickness, and so overall the capacitance decreases by a factor of two.

This result seems quite promising, but the effects of scaling on resistance are not so good. As the cross-sectional area of the wires is reduced it becomes increasingly difficult for the current to flow through the wires, and so the resistance increases. If both the thickness and width of the interconnects are scaled, then the cross-sectional area is reduced by a factor of four and the resistance increases accordingly. However, since the resistance is also proportional to the length of the wire, we find that overall the resistance increases by a factor of two. Since the resistance increases by the same factor as the capacitance decreases, the time delay stays the same as in the unscaled system. This is bad news. Although the switching times of the individual devices are reduced by scaling, the delays in propagating signals between them are not. This suggests that the speed of an integrated circuit will ultimately be limited by the need for devices to communicate with one another.

There are other more serious penalties involved in reducing the dimensions of the interconnects. Let us consider an analogy of passing water through a hose pipe. If we reduce the bore of the hose pipe it becomes more difficult to pass water through the hose at the same rate as before. In fact, if the hose is not strong enough it may even burst. A similar effect can occur when electrons are forced through a narrow wire. We have seen that the current flowing between the devices is reduced by a factor of two, but the cross-sectional area of the wires is reduced by a factor of four. As a result the current per unit area (or current density) is twice as large as it was before scaling. Since the decrease in the cross-sectional area also leads to an increase in resistance, this in turn means that more heat is generated as the current flows through the wire. The heat causes the ions to vibrate rapidly, and since they are under constant bombardment from the concentrated flow of electrons, they may literally be wrenched out of their lattice sites. This movement of the ions, known as electromigration, may be so severe that all of the atoms in a small region are removed, physically destroying the connection. Since there is no way of repairing such a tiny wire, this will render the entire chip useless.

This assessment of the effects of scaling the interconnects may seem pessimistic, but in reality the situation can be much worse. We assumed in our argument that the length of the interconnect scales down in proportion to the other dimensions, but this is not necessarily the case. In particular, although great efforts are taken in the layout of the circuits, there is always a requirement for some long interconnects which pass from one side of the chip to the other. Obviously, these are not reduced in length as the size of the devices is reduced. If anything the trend to increase the size of the individual circuits has tended to make these interconnects progressively longer. This poses a serious problem. If we repeat the above analysis but consider the interconnect length to remain the same with scaling, then we find that reducing the dimensions by a factor of two means that the capacitance remains the same, while the resistance increases by a factor of four. In consequence, the time taken for a signal to propagate along these long interconnects will actually increase quite dramatically with scaling.

We arrive at the conclusion that putting too many devices on a chip may seriously slow down the operating speed of the system. What can be done about this? One solution is to use less aggressive scaling on the interconnects. For example, suppose that the depth and width of the interconnects are left unchanged. This means that the resistance of the interconnect does not increase even when the length of the interconnect remains the same. The effects on the capacitance are not so favourable, but this may be offset by increasing the thickness of the oxide layer. Of course, if the widths of the interconnects are not scaled, then as the dimensions of the devices are reduced the interconnects take up more of the surface area. In fact, a high-density VLSI circuit of area one square centimetre will typically contain a total length of about twenty metres of interconnect. If we assume that the interconnects are, say, five microns wide, then a quick calculation shows that the interconnects cover virtually the entire surface area of the chip. To avoid this situation many VLSI chips employ several levels of wiring, each separated from the other by a layer of oxide.

A further possibility for avoiding the problems associated with interconnects is to reduce the temperature. In this way the operating temperature of the system can be thought of as another scaling parameter (although it does not follow that the same scaling factor should be applied). While it may not be feasible to control the temperature in all applications, cooling systems are already commonplace in many supercomputers. The effort is worth it because the use of lower temperatures produces a multitude of benefits. For instance, the resistance of the metal interconnects decreases, thereby reducing the time delay. The vibrations of the ions are also reduced, so lowering the susceptibility of the wires to electromigration. In addition, the problems with scaling the gate voltage can be alleviated because there is less thermal noise, and, as we shall see in the next chapter, the devices themselves operate faster at low temperatures.

We have seen then that there are many potential problems to be faced as attempts are made to increase the density of electronic devices in an integrated circuit. Nevertheless, the simple approach of scaling should allow the trend for miniaturization to continue at least into the next century, by which point it should be possible to place nearly a billion devices on a single chip. What then? Scaling theory suggests that the minimum gate length is about a fifth of a micron, but even on these length scales we have not encountered any fundamental limits. This suggests that further progress should still be possible. However, it may require a quite different technology from the one which has brought us this far.

6

Upwardly Mobile
or
How to Make Electrons Travel Faster

HOW do we make a transistor faster? In other words, what can we do to reduce the time it takes to switch it on and then off again? We have already seen that the switching time must be long enough to allow a typical electron to get from one side of the gate region to the other. The simplest way to decrease this time is to reduce the distance that the electron has to travel. This is the route we followed in the previous chapter. An alternative solution is to encourage the electrons to travel faster. This requires a far more subtle approach. To see how we can achieve this we need to have an understanding of how an electron behaves as it makes its way through the maze of atoms in a crystal.

Let us begin with a very primitive model of a crystal in which each atom is represented by a hard sphere. These spheres occupy fixed positions so that they form a regular three-dimensional grid. Figure 6.1 shows a typical path of an electron as it moves through this structure. The electron travels along a straight path until it collides with one of the spheres. Following the collision it bounces off in a new direction and continues along this course until it undergoes a further collision. The process continues with the electron ricocheting off one atom after another. This motion could well be described as chaotic. What a physicist means by chaotic behaviour is that a small difference in the initial conditions leads to vastly different results. This clearly describes the motion of an individual electron

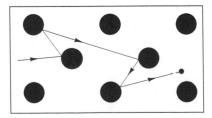

Figure 6.1 An electron, pictured as a small ball, following a complex trajectory through a crystal.

in this system. Although we could, in principle, work out exactly the new motion of the electron after each collision, a very small change in the initial direction of motion of the electron produces a quite different path after only a few collisions. The motion of the electrons as a collective body is also typical of a system that any non-scientist would call chaotic, since if we could take a snapshot of the system at any moment in time, we would find the electrons moving in a haphazard, disorderly manner with apparently no preferential direction of motion.

Suppose that a voltage is now applied to the crystal. We know from experience that this causes a current to flow. Electrons enter the crystal at the negative terminal and flow out through the positive terminal. How can this happen if the conduction electrons in the crystal are all in random motion?

These two apparently contradictory statements can easily be resolved by considering the motion of a ball in a pinball machine. Although the ball rattles around the table following a complex random path, sooner or later it ends up at the bottom of the table. This is because the force of gravity is always pulling the ball downwards, so that although the ball rebounds in random directions after each collision, it follows a curved path rather than a straight one. For example, if at some stage the ball ends up moving up the table as the result of a collision, the force of gravity will cause it to come to rest and then to travel back down the slope even if it does not undergo a further collision. Similarly, for the case of an electron in a crystal, we can think of the energy bands being tilted by the applied voltage so that the electron always ends up at the lower, positive end.

Consequently, we can see that when an electric current flows in a material, the electrons do not all move in the direction of the current as we might expect. Instead, only a very small percentage of the electrons will be moving in this direction at any given time. In order for the electrons to get from one side of the crystal to the other they follow a highly convoluted path, travelling in total many thousands of times further than the distance in a straight line. As a result, although the electrons are moving at an average speed of about a hundred thousand metres per second, they will typically move only a few metres along the direction of the current in one second.

This simple model also gives us an explanation of the origin of electrical resistance. At each collision some of the kinetic energy of the electrons is converted into heat energy. Following a collision, the electron is accelerated, gains more kinetic energy, and then loses it again at the next collision. Thus, an electric current flowing through a material produces heat. This is exactly what we observe in practice: it is the principle used in electric kettles, bar fires and light bulbs.

Armed with this basic understanding of the motion of an electron in a crystal, let us now turn our attention to the problem of reducing the time it takes an electron to travel a given distance. We can gain some insight into this by taking the analogy with the pinball table a stage further.

One way to reduce the time taken for the pinball to travel down the table is to raise the back legs of the machine, thereby increasing the tilt of the table. Since gravity now plays a more significant role a ball travelling down the table is accelerated more rapidly, while one venturing up the table is turned round in a shorter time. The application of this effect to the electron is quite straightforward. Since the degree of tilt in the crystal is determined by the strength of the electric field, an analogous effect is obtained by simply applying a larger voltage. Although this allows the transistor to switch faster, as we have seen in the previous chapter, the penalty is an increase in the amount of heat produced.

If we want to avoid an increase in heat, then rather than thinking about the speed at which the electrons travel, we should be concerned about a quantity called mobility. We can think of the mobility as representing the ease with which an electron can move through the crystal. Strictly speaking it actually tells us how quickly an electron can get from A to B for a given electric field. What we should really be doing then is trying to increase the mobility. Such an effect is easy to achieve on the pinball table: we simply remove some of the pins. In this way the ball does not necessarily go any faster, but, because there are fewer obstacles with which to collide, the total distance travelled by the ball is reduced. Consequently, the ball takes correspondingly less time to traverse the length of the table. The difficulty is, how do we relate this to the crystal?

To answer this question we first need to improve our model describing the motion of the electrons in the crystal. So far we have assumed that the atoms have no particular influence on the conduction electrons— they are just objects that get in the way. However, this is far from the truth. Each atom consists of a very small positively charged nucleus surrounded by a large cloud of electrons. As a result there are strong forces between the electrically charged particles in the atom and a nearby electron. This suggests that a quite different view is in order. Rather than picturing a blundering electron crashing its way through the network of atoms, it is much better to imagine a sensitive electron working in harmony with the crystal. As it approaches each atom it is gently deflected by the surrounding cloud of electrons. The rather surprising outcome from such a scheme is that in a perfect crystal the repulsive forces from each of the atoms actually guides a free electron through the spaces between the atoms without a single collision. This is shown schematically in Figure 6.2. This seems to be quite a fantastic feat. One way to visualize the motion of the electron is to think of a car at the top of a circular multi-storey car park. To steer a car down the spiralling roadway takes some considerable degree of concentration and would no doubt have to be undertaken at slow speed. However, if we were initially to

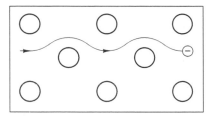

Figure 6.2 A negatively charged electron moving through a regular grid of atoms.

set the steering lock correctly, then we could just let the car roll down on its own without incident.

There is one major problem with this result. If there are no collisions, then an electron travelling through the crystal would never lose any kinetic energy. Consequently, the material should offer no resistance to a flow of electrons! How do we reconcile this with our experience that all materials (with the exception of superconductors) do exhibit resistance? The answer relies on one simple word. We assumed the crystal to be perfect, a word which in this context implies that all the atoms are identical and occupy positions in a regular three-dimensional grid. Such a structure can exist only in the mind of the theorist. In any real structure we always have to contend with one major alteration to this picture. The atoms simply do not sit still.

Let us imagine what would happen if we arranged a number of chairs in a series of regularly spaced rows. We then assign a five-year-old child to sit in each seat. Although we make sure that they do not get up or change seats, the children most certainly will not be stationary. Some will be leaning across to talk to their neighbours, other may be looking to see who is behind them, or they may just be fidgeting about. We would see just the same picture if we could observe the atoms in a crystal at room temperature. Each atom stays close to its assigned position in the crystal, but every one is in constant motion. Consequently, the atoms are never quite where they ought to be. Imagine how an electron reacts as it travels through this rabble. Rather than gliding undisturbed between the atoms, the electron is constantly jostled by their movements. Every so often it comes rather too close to an atom and receives an extra-large nudge so that it is launched on to a new course in a different direction. The result of many such encounters is very similar to that of our first model. The electron ends up following a hideously complex route accompanied by frequent changes of direction as it attempts to make its way through the network of atoms.

The difference with this new model is that it is now possible to see how to improve the situation. The degree of motion of the atoms, or in other words the average distance by which they stray from their assigned positions, depends on the temperature of the solid. The higher the temperature, the more energy is available for the atoms to move about. Conversely, if the solid is cooled down, then the atoms move about more sluggishly and stay closer to their ideal positions. This reduces the chance of the electron being knocked off course. One way to improve the mobility is therefore to lower the temperature. At minus one hundred Celsius the mobility of the electrons is much greater than that at room temperature. It would seem that the further the temperature is reduced, the more the mobility increases. Unfortunately, this is not true, at least as far as semiconductors are concerned. Lowering the temperature to minus two hundred Celsius may even cause the mobility to decrease. Why is this?

There are of course other reasons why a crystal is not perfect. In particular, no material is ever one hundred percent pure. All crystals are contaminated to some extent by unwanted impurity atoms. These atoms may be slightly larger or slightly smaller than the other atoms in the crystal. In either case, their presence causes a disturbance in the crystal structure over a small, localized region, which affects the motion of any nearby electrons. The best way of reducing the effects of these impurities is to strive for ever increasing degrees of purity in the crystals. As we have seen in Chapter 2, exemplary efforts in this direction with semiconductor wafers have reached the level where less than one atom in every ten billion is an accidental impurity.

End of story, or so you might think. Unfortunately, some impurities are essential in semiconductors if we are to construct transistors. We must intentionally add impurities to the crystal, in the form of acceptors or donors, in order to control the conductivity of the material. Let us consider the example of introducing phosphorus donor atoms into a silicon crystal to produce n-type silicon. Remember (from Chapter 2) that the phosphorus atom has one more valence electron than silicon. When the electron is removed to become a conduction electron we are left with a positively charged ion. The attractive force exerted by this positive ion on the negatively charged conduction electrons extends over a considerable range. Not only the electrons in the immediate vicinity of the ion, but also those which are far from the ion, are deflected by it. The course followed by a typical electron is illustrated schematically in Figure 6.3. One way to picture the effect is to imagine the electron travelling across a relatively flat

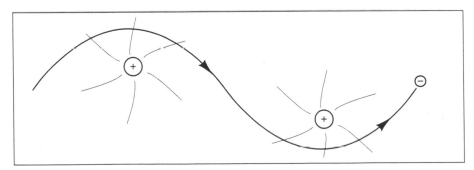

Figure 6.3 An electron is strongly deflected by any charged impurities.

surface which is pock-marked by dimples at the positions of the phosphorus ions. The surface is then rather like one of those old-style upholstered settees which have buttons to pull the filling together in the cushions. Although the buttons are few and far between, the depression produced by each button extends over a large region. If a marble is rolled across the surface, it does not follow a straight line, but instead deviates from its course each time it encounters one of these depressions. Since the depressions extend over such a large area, even a small number significantly affect the motion of the marble.

We have therefore arrived at what appears to be a Catch-22 situation: the impurities that are introduced into the crystal to provide the much needed supply of conduction electrons are the very same ones that restrict the mobility of these electrons. There is a rather elegant solution to this conundrum, but before we can provide the answer we must first take a look at some other semiconducting materials.

So far we have concentrated exclusively on silicon. This is not without good reason since silicon has been, and still is, used for the vast majority of commercially available semiconductor devices. It is easy to see why silicon has achieved such a dominant position in the marketplace by taking a look at some of its attributes. It is an extremely abundant material—indeed, over a quarter of the Earth's crust is composed of silicon atoms in one form or another. Consequently, it is very cheap. It is also a good conductor of heat, a factor which is of great importance when considering how closely individual transistors can be packed together in an integrated circuit. The other most relevant property is that it possesses a native oxide which is an excellent insulator. As we have seen, this feature is exploited in the MOSFET and in isolating the interconnecting wires on the surface of a silicon chip from the transistors below.

Despite the apparent monopoly held by silicon, other semiconductor materials have characteristics which favour their use in certain circumstances. One such material, which has been used for many years in the manufacture of high-speed transistors, is gallium arsenide.

Gallium arsenide is composed of two different elements, gallium and arsenic. The gallium atoms have only three valence electrons, and therefore have room for five more; the arsenic atoms conveniently have five valence electrons, and can accommodate three more. This seems promising — if we combine gallium and arsenic atoms in equal quantities we have an average of four electrons per atom, as in silicon. However, the situation is more complicated than in silicon since the arsenic atoms have a tendency to steal electrons from the gallium atoms. Consequently, we end up with negatively charged arsenic ions and positively charged gallium atoms. We have seen an extreme example of such an effect in Chapter 1 when we considered sodium chloride. As in sodium chloride, the ions arrange themselves so as to neutralize the electrical charge on as small a scale as possible. Each negative arsenic ion joins up with four positive gallium ions, and

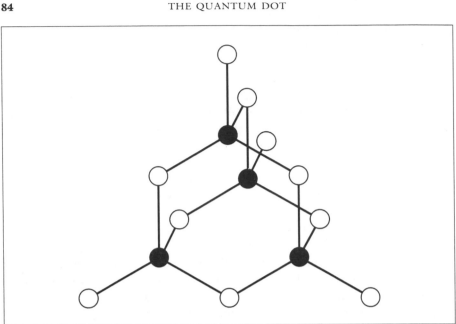

Figure 6.4 The structure of a gallium arsenide crystal. The black and white spheres represent gallium and arsenic atoms, respectively.

each gallium ion bonds with four arsenic ions. We therefore end up with an arrangement in which the two types of atoms alternate. The resultant structure is shown in Figure 6.4. Comparison with Figure 1.5 shows that the structure resembles that of diamond, and therefore also that of silicon. The similarity with silicon extends to several other properties. For example, there are precisely enough electrons to fill the valence band, and so the material is a semiconductor—it has a band gap slightly larger than that of silicon. Similarly, the conductivity can be controlled by doping, although we have the rather strange situation in which silicon atoms can act either as donors or as acceptors depending on whether they replace gallium or arsenic atoms, respectively. Alternatively, elements with six valence electrons, such as selenium, form donors if placed on an arsenic site, whilst atoms such as zinc, with only two valence electrons, will accept another electron if placed on a gallium site.

Why is gallium arsenide superior to silicon for use in high-speed transistors? The reason is attributable to two rather subtle differences between the materials. Firstly, suppose that we take two similar-sized samples, one of gallium arsenide and the other of silicon, and apply the same voltage to each. The tilt produced by the voltage is the same in each case, and yet the conduction electrons in the gallium arsenide are found to be accelerated much more rapidly between collisions than those in the silicon. How do we explain this behaviour? Does this mean that the electrons in a crystal of gallium arsenide are intrinsically different from those in silicon? Of course not: electrons are the same whatever atom or crystal they belong to. A better way to think of this is to consider taking two identical car engines and placing one in a large family saloon and the other in a lightweight sports car. When coupled to the sports car chassis the engine is capable of producing much faster acceleration than when placed in the saloon. However, the engines are still the same: it is what we put them into that matters. Similarly, electrons in gallium arsenide respond differently to those in silicon because of differences between the two crystals. If we were to pluck an electron out of a silicon crystal and place it into a crystal of gallium arsenide, it would behave just like all the other electrons in the gallium arsenide. We refer to this property by saying that an electron has an effective mass in a given material, and so the effective mass of an electron in gallium arsenide is less than that of an electron in silicon.

To explain the second difference between the two materials we need to make a further revision to our model of electron motion in a crystal. Back in Chapter 2 we introduced the idea of the valence band in a crystal forming as a result of a large, but finite, number of very closely spaced energy levels, each of which has the capacity to hold two electrons. In this way we discovered that the valence band in a semiconductor is full. When talking about the conduction band we have never had to worry about the presence of individual levels since this band is always comparatively empty. Nevertheless, we need to remember that the conduction band is also a collection of discrete states. More to the point, only a very limited number of these will be accessible to electrons with energies near the bottom of the band. Let us see then how the presence of these levels affects the electron as it travels through the crystal. For simplicity, we will assume that the electron moves in only two dimensions. Furthermore, we will associate each of the levels with a particular direction of motion of the electron. Suppose that in a certain material there are only four such levels within a range of energy accessible to the electron, corresponding to the electron moving down the slope, up the slope, or either right or left across the slope. Initially the electron moves down the slope, making a bee-line for the positive terminal. The only way in which the electron can be knocked off this course is if it receives a nudge large enough to change its direction of motion by ninety degrees. This we can consider to be a rather improbable event. If we make a comparison with a material in which there are twenty different levels, and so twenty different possible directions of motion, then there is a far greater chance of the electron being diverted from its course into a different state. A similar argument applies in a real crystal; the probability that an electron changes state depends on the number of states available. It turns out that in silicon there are many more conduction states available to electrons near the bottom of the conduction band than there are in gallium arsenide. In technical parlance we say that the density of states is larger in silicon than in gallium arsenide. The consequence of this is that an electron travelling through a crystal of silicon has a greater chance of being knocked off course than a similar electron in gallium arsenide.

In actual fact, these two effects are closely related to one another. In general, we can say that a small effective mass is indicative of the fact that there are relatively few energy levels near the bottom of the conduction band. However, for the moment our main concern is that the mobility of an electron in gallium arsenide is increased by a factor of six compared to silicon. This translates directly into an increase in speed of the transistor. If we make two transistors of equal size, one from silicon and one from gallium arsenide, then under comparable conditions the gallium arsenide one should be capable of operating approximately six times as fast as the silicon one.

Unfortunately, it is not quite this simple in practice. Much of the technology used in fabricating silicon chips cannot be transferred directly to gallium arsenide. For instance, unlike silicon, gallium arsenide does not possess a suitable native oxide. This means that alternative methods must be introduced both in the production and in the design of the transistors. Furthermore, gallium arsenide vaporizes at the temperatures traditionally used for processing silicon. Thus, the manufacturers of gallium arsenide integrated circuits have largely had to start from scratch. As a consequence, gallium arsenide integrated circuits are not at such an advanced stage of development as those constructed from silicon. The lower levels of integration, coupled with the difficulties encountered in the manufacturing process, mean that gallium arsenide circuits are far more expensive than silicon ones. Nevertheless, gallium arsenide transistors are used in cases where speed of operation is of absolute importance. These include military applications and the latest generation of supercomputers such as the Cray-3.

While the speed advantage gained by using gallium arsenide transistors represents a significant advance, the mobility is ultimately limited by the same restriction as silicon. At low temperatures it is the impurities necessary for doping the material which disrupt the motion of the electrons. What we need to do is to separate the two physically. The impurities can easily be controlled so that they are introduced into only one portion of the crystal. This principle is called modulation doping because we selectively dope certain regions of the crystal rather than uniformly doping the whole area. We then need to entice the conduction electrons into a different region. This is the difficult part. How can we achieve this? A good starting-point is to look at the family tree of gallium arsenide.

Gallium arsenide belongs to a large family of materials known as three–five semiconductors (this is usually written using Roman numerals, III–V). This name derives from the fact that gallium, with three valence electrons, falls in Group Three of the periodic table, while arsenic is in Group Five and has five valence electrons. A number of different semiconductors can be formed by substituting other Group Three atoms, such as aluminium or indium, in place of the gallium atoms, or by replacing the arsenic atoms with others from Group Five, such as phosphorus or antimony. This leads to combinations such as aluminium arsenide, gallium phosphide and so on, all of which have the same crystal structure as gallium arsenide, but rather different electrical and optical properties.

An even greater range of materials, with intermediate properties, can be obtained by forming alloys. Alloying is a term which we usually associate with metals, where it means that the material is a mixture of two elements. A common example is brass, which is an alloy of copper and zinc. When referring to III–V semiconductors, alloying means that either the Group Three or the Group Five atoms are a mixture of two different elements. As an example, suppose we choose to replace some of the gallium atoms in a crystal of gallium arsenide with atoms of aluminium. We therefore have a crystal where alternate atoms are arsenic, and the gallium and aluminium atoms are randomly assigned to the other positions. This material goes by the name aluminium gallium arsenide, which is rather a mouthful. Consequently, it is often simply referred to as 'algas', a term derived from its chemical representation, AlGaAs. A very useful characteristic of this alloy is that the aluminium atoms are almost exactly the same size as the gallium ones that they replace. This means that the atomic spacing in the alloy is almost identical to that in the gallium arsenide. Consequently, it is possible to make a single crystal which is composed of algas in one region and pure gallium arsenide in another.

It is worth being side-tracked for a moment to examine how such a structure is produced. The most common methods involve the use of a vapour to supply atoms which then attach themselves to the surface of an existing crystal in such a way that they continue the ordered pattern already present in the crystal. In this way the crystal appears to 'grow' as successive layers of atoms are built up. Suppose we start with a crystal of algas. We can grow additional layers of the crystal by using three gas supplies in which the molecules contain arsenic, gallium and aluminium atoms, respectively. The algas crystal is kept at a temperature of a few hundred Celsius—well below the point at which the algas vaporizes, but high enough to decompose any of the gas molecules which come in close contact with the surface so that they leave behind the Group Three or Group Five atoms to add to the crystal. If we now wish for successive layers of the crystal to be pure gallium arsenide, we simply turn off the gas supply containing the aluminium. At any later stage we can turn this back on if we want to revert to algas again. This process is called chemical vapour deposition, or CVD. The rate at which the crystal grows is extremely slow, typically in the range of ten to one hundred atomic layers per second. At this rate it would take about a day to grow a crystal one millimetre thick. However, the technique is designed for growing

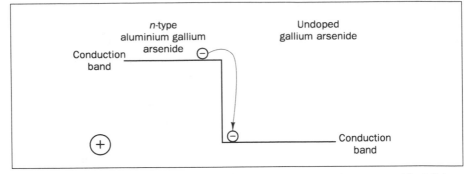

Figure 6.5 Conduction electrons originally in the aluminium gallium arsenide fall into lower energy states in the gallium arsenide. As a result, positively charged ions are left in the aluminium gallium arsenide.

layers typically less than a thousandth of a millimetre in thickness, and can produce a crystal which changes from algas to gallium arsenide over a distance of only a few atomic layers.

What do we achieve by producing such a structure? Although algas is very similar in many ways to gallium arsenide, there is one important difference. The amount of energy that is required to raise an electron from the valence band into the conduction band increases with the concentration of aluminium in the alloy. As a result, the conduction bands in the two materials do not line up: the lowest energy states for conduction electrons in the gallium arsenide are considerably below those in the algas region. This situation is shown in Figure 6.5. Let us consider a conduction electron in the algas region of the crystal. The conduction band is fairly flat, and we can imagine the electron wandering about at random. Sooner or later, it comes close to the gallium arsenide region, tumbles down the steep cliff, and ends up in the conduction band of the gallium arsenide. This is a one-way process—once the electron has fallen into the lower energy levels in the gallium arsenide, it no longer has the energy to get back into the algas layer.

This is just what we have been looking for. We have discovered a way out of the Catch-22 problem that we faced earlier. If the donor atoms are placed in the algas region and the gallium arsenide is kept as free from impurities as possible, then the conduction electrons tend to collect in the gallium arsenide region where they have much higher mobility. The electrons which disappear over into the gallium arsenide region leave behind positively charged ions. These ions exert an attractive force on the conduction electrons and try to pull them back into the algas layer. However, once the electrons have fallen into the lower energy gallium arsenide states they are unable to gain sufficient energy to move back up to the algas conduction band. The electrons are therefore trapped in a narrow layer of the gallium arsenide close to the interface with the algas. 'Trapped' is

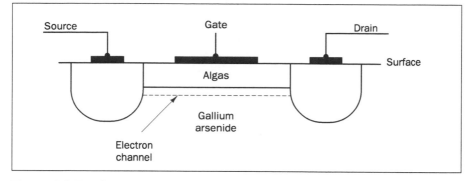

Figure 6.6 The formation of a high mobility electron channel in a modulation doped FET. Note the similarity to the MOSFET in Figure 5.1.

not quite the right word, since although they cannot move either back into the algas region or further into the gallium arsenide, they are free to move along a direction parallel to the interface, unhindered by the impurity atoms.

The most obvious way to take advantage of this behaviour is to design a device in which the flow of electrons is parallel to the interface. Figure 6.6 shows how this can be achieved in a structure which is reminiscent of a MOSFET. A source and a drain are added at opposite ends and a metal gate contact is made to the top of the algas layer. The thin layer of gallium arsenide at the interface then forms the channel for the electrons to travel through the gate region. By altering the voltage applied to the gate the number of electrons in the channel can be varied, and so the device can be switched on or off. Due to the enhanced mobility of the electrons, the switching times are typically ten times smaller than can be achieved with a standard silicon MOSFET. Such devices have been produced by many research establishments under various names, such as MODFET for modulation doped field effect transistor, or HEMT for high electron mobility transistor.

So far we have concentrated entirely on transistors in which electrons carry the signal. Whilst it would be possible to build a computer using only these n-channel FETS, it is highly desirable to be able to build similar p-channel FETs in which the signal is carried by holes. In particular, in Chapter 4 we introduced a particularly favourable arrangement called complementary logic in which n- and p-channel transistors are used in pairs. In such an arrangement there is clearly little point in speeding up the n-channel transistors if the p-channel transistors remain much slower. Accordingly, we should also attempt to increase the mobility of the holes.

The motion of the holes through a crystal is entirely analogous to that of the electrons, except that the holes tend to occupy the highest energy levels. In this

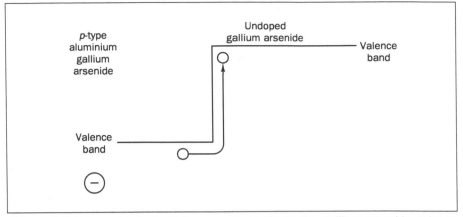

Figure 6.7 Holes migrate to higher energy levels in the gallium arsenide valence band.

respect we can think of the holes as behaving like air bubbles trapped beneath a surface. Consequently, when we tilt the energy bands by applying a voltage, the holes travel towards the higher, negative end. Just like the electrons, holes are deflected from their course by the movement of the atoms in the crystal and by impurities. We can therefore apply the same modulation doping technique that we employed to make the electrons travel faster. Since the energy states available in the valence band of the gallium arsenide are higher than in the algas (see Figure 6.7), we again place the dopant atoms in the algas region and leave the gallium arsenide layers undoped. The holes diffuse away from the acceptors, and on crossing into the gallium arsenide region they take advantage of the higher energy valence states which are found there. Again, the holes are attracted towards the algas layer by the presence of the negatively charged ions, but are unable to return because of the energy difference between the layers. Thus, we can construct a p-channel modulation-doped FET along similar lines to the n-channel device.

Unfortunately, this still does not make the p-channel FETs as fast as their n-channel counterparts. The holes find it more difficult to move through the gallium arsenide than the conduction electrons. In particular, they are not accelerated as rapidly by an electric field. For the electrons in different materials we described this property by referring to an effective mass of the electrons. If the holes are accelerated more slowly than the electrons then it follows that they must have a larger effective mass than the electrons. (It may seem a rather strange concept to associate any mass with a hole, but we should remember that the effective mass is really a property of the crystal.) It turns out that there are actually two types of holes, with different effective masses. They are referred to as light holes and heavy holes. The light holes have a considerably smaller effective mass, usually comparable with that of the electron, and are therefore accelerated more rapidly by an electric field than the heavy holes.

We can best visualize how the two types of hole affect the switching speed of a transistor by analogy with a short story (although cynics might describe it as a tall story!). Our tale is based around two communities, Sourcetown and Drain City. They are about a hundred and fifty kilometres apart and are linked by a good motorway. Between them they devise a method for sending coded messages. If Sourcetown wishes to send a 'yes' signal, then a car is sent to drive the one hundred and fifty kilometres to Drain City. If the signal is 'no', then the car is not sent. Correspondingly, Drain City interpret the arrival of the car as a 'yes', and the non-appearance as a 'no'. However, it is soon realized that it is not sufficient to send just one car. Suppose that it breaks down on the way. Then what was initially a 'yes' would be interpreted as a 'no'. A more reliable system is decided on whereby one hundred vehicles are sent out from Sourcetown. These vehicles turn out to be an assortment ranging from high-performance cars, capable of covering the distance in a mere forty-five minutes, to a number of slow lorries which take up to three hours to complete the journey. At Drain City they monitor the number of vehicles arriving over a series of three-hour periods. If a large number of vehicles arrive in one period they know that a 'yes' has been sent, and if very few arrive they interpret this as a 'no'. One day, an important official in Sourcetown decides that it is necessary to send the signals more frequently than one every three hours. He instructs Drain City that they will now be transmitting a signal every hour instead. At Sourcetown they begin by sending a 'yes', and during the first hour at Drain City they record the arrival of almost all the high-performance cars. This is a lot less than the total number of vehicles they were expecting, but enough for them to assume that the signal sent was a 'yes'. In the next hour, Sourcetown transmits a 'no', and so no vehicles are sent out. However, there are still many lorries out on the road, and some of these arrive throughout the next hour at Drain City. Again, there are fewer vehicles than expected for a positive response, but just as many as in the previous hour, therefore this signal is erroneously interpreted as another 'yes'.

In this example it is obviously the lorries which limit the speed at which signals can be sent. One solution would be for the counters at Drain City simply to ignore the lorries and count only the number of fast cars arriving. In a similar way, it is the heavy holes which determine the switching speed of a transistor. Unfortunately, it is not possible to design a transistor which registers only the arrival of the light holes. When a signal is received at the drain of a p-channel transistor we have no way of knowing whether it has been carried by light holes or heavy holes, or both. We therefore have to make the switching time long enough to allow both the light and heavy holes to traverse the gate region.

An alternative solution to the problem would be for Sourcetown to send only the fast cars. The application of this approach seems more promising. Although we cannot distinguish the light holes from the heavy holes as they arrive at the drain, there is a way in which we can ensure that almost all the carriers of the signal are light holes. In gallium arsenide both light and heavy holes at the very top of the valence band have exactly the same energy, and so both types of hole exist. As we discussed previously for the electron, the effective mass of a particle is closely connected to the number of energy states available to that particle. Consequently, we find that there are far fewer states for light holes than for heavy holes, and so heavy holes are quite substantially in the majority. We can visualise this by referring to Figure 6.8(a) which shows schematically the situation at the top of the valence band. Here we use circles to indicate the number of states available at corresponding values of energy. The light hole and heavy hole states are shown separately for clarity and the dashed line represents the Fermi energy. We assume that the situation is described by the idealized condition described in Chapter 2, i.e. all the states below this level are occupied by electrons and all the states above it are vacant.

In order to construct a high-speed p-channel transistor we must somehow ensure that the light holes outnumber the heavy holes. This could be achieved if we could somehow push the light hole states up to higher energies in comparison to the heavy hole states, as in Figure 6.8(b). Since the Fermi energy is now greater than the highest energy states available to heavy holes, this means that all of these states are occupied by electrons, and so there are no heavy holes. The only vacant positions in the valence band now correspond to light hole states.

It is all very well to propose increasing the energy of the light hole states, but how do we achieve this in practice? The answer is quite simple; if we apply pressure to the crystal the light and heavy hole states respond in different ways. It is therefore possible to achieve the situation shown in Figure 6.8(b) by applying an external force to the crystal. However, an even better solution is to use the forces between the atoms in the crystal to create a localised pressure in a small part of the crystal. Let us see how this is achieved.

Suppose we introduce a different Group Three element, indium, into the gallium arsenide crystal. Some of the gallium atoms are replaced by indium atoms and the resulting alloy is called indium gallium arsenide, or 'ingas' for short. The introduction of a different atom alters the band gap energy of the crystal as we discovered with algas. In this case, the addition of indium causes the band gap energy to be reduced in comparison to gallium arsenide. A second change is to the crystal structure itself. Unlike the algas crystal, the indium atoms are significantly larger than the gallium atoms they replace. Ideally, then, the atoms in indium gallium arsenide would space themselves further apart than in gallium arsenide in order to make room for the larger indium atoms. This is certainly true if we compare the atomic spacings in two such crystals. However, if we start

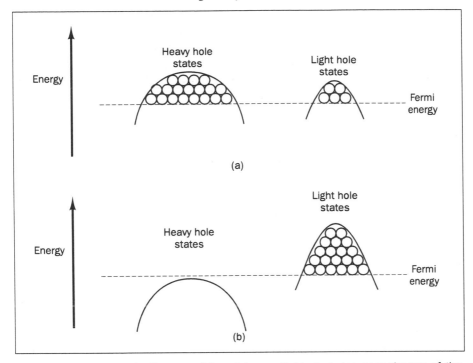

Figure 6.8 A schematic diagram of the states available to holes near the top of the valence band. In this picture the holes are represented by circles and it is assumed that all the states above the Fermi level are occupied by holes and all those below the Fermi energy are occupied by electrons. In (a), which represents the situation in most crystals, the highest energy is the same for both the heavy hole and the light hole bands. In (b), which represents a compressively strained crystal, there are higher energy states available in the light hole band, and so all of the holes are light holes.

to grow indium gallium arsenide on top of an existing gallium arsenide crystal the results are somewhat different: the new layers of indium gallium arsenide accept the atomic spacing already laid down by the gallium arsenide crystal . This means that the atoms are squashed together more closely than they would otherwise be in this plane. This squashing of the atoms is referred to as compressive strain, and has virtually the same effect as if we had applied external pressure to the crystal. The important side-effect for the present purpose is that the light holes are pushed up in energy, whilst the heavy holes are pushed down. Since the ingas crystal has a smaller band gap energy than gallium arsenide, we can now place the acceptor dopant atoms in the gallium arsenide and the holes will accumulate in the ingas layer. Furthermore, as we have just seen, these holes will occupy the states which have a very small effective mass. Consequently, if we use this system in an FET, only the light holes will play an important part, and the resultant p-channel device can be made to operate at approximately the same speed as an n-channel one.

It may seem that we have gone to quite fantastic lengths in order to increase the mobility of the electrons and holes. However, the incentive to produce ever faster devices is so great that even this is insufficient. To this end we will consider one further refinement.

Our main aim throughout this chapter has been to eliminate collisions. We have done our best to ensure that when an electron (or hole—similar principles apply) enters the gate region it travels across to the other side with the minimum deviation from its course. Let us look at what happens to the velocity of such an electron as it travels along the channel through the gate region. It enters with virtually zero velocity and is accelerated down the slope created by the voltage like a miniature drag racer, its velocity increasing the further it travels. Of course, even if it suffers no collisions, it cannot go on accelerating indefinitely. Just as a car will reach a maximum speed even if the accelerator is kept pressed to the floor, so the electron reaches a maximum velocity after a short time. For an electron in a semiconductor, this maximum velocity is typically about two hundred kilometres per second, or roughly one thousandth of the speed of light. However, the time taken for the electron to traverse the gate region depends not on the maximum speed but on the average speed of the electron over the whole distance. Since it sets off from what is virtually a standing start, the average velocity is likely to be considerably less than the maximum velocity. In a very short channel the electron may not even travel far enough to reach the maximum velocity. The way around this is to give the electron a flying start. If the electron could be fired into the gate so that it is already travelling at close to its maximum velocity as it enters the channel, then the average velocity would be virtually the same as the maximum velocity. This is called ballistic transport since the electron in this case behaves like a missile, travelling at a constant speed throughout.

How do we succeed in launching an electron into the gate at such speed? We have already described a suitable structure without realizing it. In the modula-tion-doped FET we said that the conduction electrons produced in the algas layer fell into the lower energy gallium arsenide conduction band and were then used for carrying a current parallel to the interface. What would happen if we turned the device around so that the electrons travel at right angles to the interface? As the electron moves from the algas into the gallium arsenide region it behaves like a ball falling from a cliff top. It gains speed very rapidly, much more so than when it rolls down the gentle slope produced by the externally applied potential. Consequently, if the cliff is high enough, the electron attains its maximum velocity almost instantaneously.

The devices we have described in this and the previous chapter represent the limits of what we will refer to as conventional microelectronics. We have reduced the dimensions of the individual transistors approximately ten thousandfold compared to discrete devices, and have then tweaked the structures to maximize the mobility of the electrons and holes. We have considered various strategies in order to achieve this. We have looked at different types of material—gallium arsenide, for example, is found to be inherently capable of producing much faster devices than silicon. Modulation doping has been explored and we have shown how it can be used to obtain exceptionally high mobilities, and even the possibility of firing the electrons ballistically through the transistor has been addressed. However, although there are many small differences between these structures, the basic principles governing the operation of the transistor have remained essentially the same throughout. To push back the boundaries even further we not only need to look at new types of transistor, but we have also to consider a new branch of physics. We must now enter the strange and often deceiving world of quantum theory.

When is a Particle not a Particle?
The Importance of Electron Waves

WE have already encountered several aspects of quantum theory: the presence of discrete energy levels in atoms; the exclusion principle which tells us how these levels are occupied; and the formation of energy bands in solids. We did not make a big deal of it at the time since, once we accepted the premise of the band gap in semiconductors, we were able to understand the operation of the transistor without worrying about whether we were using results of quantum theory. However, if we wish to probe further—to understand the devices of the future, those under development, and some which are already commercially available—then there is no escape. We have to confront quantum theory head on.

We will not be concerned with any of the deep philosophical questions which arise from quantum theory. These are addressed in many other books, and do not concern us here. Instead, we will concentrate on the practical effects—how quantum theory affects the behaviour of semiconductor devices. From this point of view, there are two principal consequences, and we will devote a chapter to each. This chapter deals with wave–particle duality. If you have not encountered this phrase before, do not worry—all will be revealed in due course. The following chapter describes a phenomenon called tunnelling. It is only after exploring these concepts that it is possible to understand the operation of the quantum transistor, a subject which we turn to in Chapter 9.

What do we mean when we say that something behaves like a wave or like a particle? Let us begin by taking light as an example. Is light best described as a wave, or does it really consist of small discrete particles? Such questions are not new: in the seventeenth century there was great debate about the matter. Isaac Newton was an advocate of the corpuscular nature of light. He reasoned that if light consists of small particles then the behaviour of a light ray can be determined by considering the forces acting on each of the particles. This model seemed particularly attractive since it bore certain similarities to Newton's highly successful treatment of the motion of the planets. An alternative, although less popular, theory was put forward by the Dutch astronomer and physicist, Christiaan Huygens, who devised a model for the propagation of light based

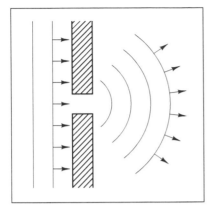

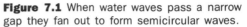

Figure 7.1 When water waves pass a narrow gap they fan out to form semicircular waves.

on a wave hypothesis. In this scheme we can picture light waves spreading outwards from a light bulb like the ripples which appear when a stone is dropped into a still pond. Which of these models is correct? The root of the problem seems to be that it is not possible to see either the particles of light or the light rays themselves. The nature of light has to be inferred by careful experimentation.

Let us concentrate on waves for a moment and ask ourselves what sort of properties we would expect to see if light is really a wavelike phenomenon. One of the simplest demonstrations of wavelike behaviour is diffraction. This occurs when a wave is partially obstructed by an obstacle, with the result that some of the wave is diverted sideways from the original direction of motion. We can illustrate this example graphically using water waves in a wave tank or shallow bath if we place two obstacles across the tank so that they form an almost complete barrier, leaving just a small gap in between. Suppose that we now make waves on one side of the barrier. Although only a small portion of the wave can pass through the gap, on the other side of the barrier we find that the wave spreads outwards in all directions. If the gap is narrow enough, the diffracted wave fans out to form an almost perfect semicircle, as shown in Figure 7.1.

A similar result can be obtained with light rays by placing an opaque sheet in the path of a beam of light, leaving just a narrow slit for the light to pass through. To observe the outcome of the experiment we place a screen at a further distance beyond the slit. As we might expect, there is a bright image of the slit in the centre of the screen, but it is noticeable that the edges of the image are quite blurred. The beam appears to have been diverted sideways by the slit, rather like the diffracted water waves, but it is not convincing proof of the wavelike qualities of light. Besides, Newton managed to predict a similar effect for a light beam consisting of discrete corpuscles.

Conclusive evidence for the wave nature of light was not discovered until

1801. Thomas Young spent most of his life as a physician rather than a physicist, and his interest in light was primarily from a physiological viewpoint. However, he is perhaps best remembered for his demonstration that light travels as a wave. Young's experiment was a simple extension of the above idea, except that he had two slits in his opaque sheet instead of one. What did he observe? Two fuzzy images? No: instead he saw a set of regularly spaced bright and dark lines. How can we explain this? To do so we need to examine the arrangement shown in Figure 7.2(a) which represents a plan view of the apparatus. We can picture the light wave spreading out until it encounters the opaque sheet. At this point each of the slits gives rise to a set of waves similar to the diffracted water waves shown in Figure 7.1. The crucial feature is that the waves spreading outwards from slit 1 interfere with those originating from slit 2, and it is this interference which gives rise to the pattern of bright and dark lines.

This is a very general description of how the interference pattern arises. If you are unfamiliar with this result it is worth while taking a more detailed look at how this image is constructed. Let us consider a point on the screen midway between the two slits, as indicated by A in Figure 7.2(b). The light arriving at this point via slit 1 has travelled precisely the same distance as that from slit 2, therefore the two sets of waves are in phase—when one is at a peak the other is also at a peak. This is illustrated on the extreme right of Figure 7.2(b). The two waves combine to form a single wave, the peak amplitude of which is larger than that of either of the constituent waves. This is confirmed by the appearance of a bright image at this point on the screen. We describe this condition as constructive interference.

At point B, which is slightly closer to slit 1 than to slit 2, the situation is quite different. The light which passes through slit 2 has to travel further than that which arrives via slit 1. Let us suppose that the distance from slit 1 to point B on the screen corresponds to exactly one million wavelengths, and that the distance from slit 2 is half a wavelength longer. Surely such a small difference can have no significant effect on the outcome. We are talking about a percentage difference of 0.00005%, or in absolute terms a distance of one four thousandth of a millimetre. However, when we examine the screen we find that it is dark at this point. What has happened? The difference of half a wavelength between the two distances means that when the light waves arrive at the screen they are out of phase. When one is at a peak, the other is at a trough. Consequently the light waves cancel each other out, and the screen appears to be unlit in this region. This is referred to as destructive interference.

Finally, let us consider point C, which is closer still to slit 1. The two light beams travel along paths of different length as before, but this time the difference is equal to a whole wavelength. This brings the two waves back in phase again, and so produces a bright region on the screen. It is a simple extension of this argument to show that bright regions occur at any point where the difference

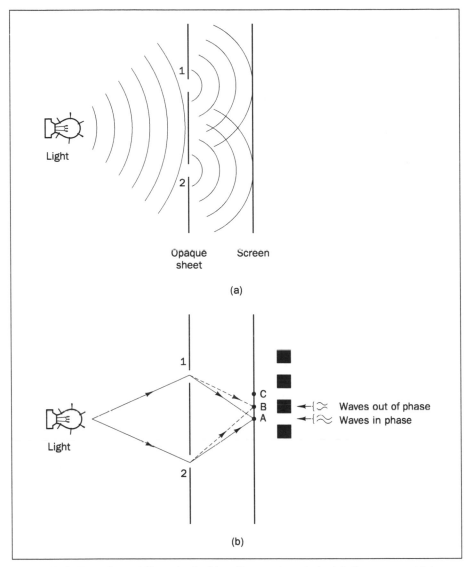

Figure 7.2 Plan view of Young's double slit experiment. In (a) the waves which are diffracted by the slits are shown as a series of semicircular wavefronts, as in Figure 7.1. This shows how the two sets of wavefronts overlap to produce interference. In (b) a ray diagram is used to indicate the paths taken by the light rays arriving at points A and B on the screen.

in the path length is approximately equal to a whole number of wavelengths. Similarly, dark regions must occur for the other cases.

In conclusion, we can see that by treating a light beam as a wave which travels through both slits simultaneously, we are able to account for the sequence of bright and dark bands observed on the screen. In comparison, the particle theory of light, which requires that each particle must pass through either one slit or the other, cannot explain these results.

As a result of this and several further experimental investigations, the acceptance of the wave nature of light rapidly became universal, and in 1864 the Scotsman James Clerk Maxwell succeeded in putting the whole thing on a solid mathematical foundation. Maxwell's theory treated visible light, and indeed all other forms of electromagnetic radiation such as X-rays and radio waves, as wavelike phenomena. Although there was some disagreement about the inter- pretation of Maxwell's theory, these matters were finally laid to rest by the German physicist Heinrich Hertz in 1888. And so it seemed that by the end of the nineteenth century physicists knew pretty much everything there was to know about light. It is just at times like this that nature decides to throw a spanner in the works.

In 1905, a young man named Albert Einstein solved a rather baffling problem concerning a phenomenon known as the photoelectric effect. Some years earlier it had been observed that electrons are emitted when ultraviolet light is shone on to a metal surface. This does not seem an altogether surprising result. We know that the valence electrons are trapped within a metal because of the attractive forces from the positive ions. We might reasonably expect, therefore, that if the energy in a light beam can be transferred to an electron near the surface of the metal, then the electron will be able to escape. However, the wave theory of light is unable to explain two important details. If an intense beam of ultraviolet light is shone on the surface, then a large number of electrons are emitted. If the light is dimmed the number of electrons decreases. However, no matter how far the intensity of the light is reduced, there is no noticeable time delay between the arrival of the light and the emission of the electrons. If we consider an area of metal measuring one square centimetre, there are about a million billion atoms on this surface. If the energy of the light wave is approximately evenly distributed between all these atoms, then it should take minutes, hours, or even days for any one electron to accumulate enough energy to enable it to escape from the metal. An even stranger result occurs if we use light of slightly longer wavelength. For example, if visible light is used in place of ultraviolet, then the emission of electrons ceases completely. This might be expected on the grounds that the visible light waves carry less energy, but we should be able to compensate for this simply by using a brighter light source. This is not the case. We find that no electrons are emitted at this longer wavelength, regardless of the intensity of the light.

Einstein found a simple way to explain this behaviour, but in doing so he contradicted the wave theory of light which had been so carefully developed over the previous hundred years. Einstein's solution was based on a notion proposed five years earlier by the German physicist Max Planck. Planck had proposed that light is emitted in small packets, or quanta, of energy. While this was certainly a revolutionary idea, it did not in any way conflict with the theory that light travels as a wave. However, in order to account for the photoelectric effect, Einstein was forced to conclude that light must actually propagate as packets of energy called photons. In this way, a photon transfers all of its energy to a single electron in the metal. This explains why electrons are emitted instantaneously even with light of low intensity. The energy of the light beam is not distributed over a large number of atoms as we supposed. Instead, the energy of each photon is concentrated into a single electron, and so there should be a one-to-one correspondence between the number of photons absorbed at the surface and the number of electrons leaving it. Following Planck's suggestion that the energy of the photon is inversely proportional to the wavelength of the light, Einstein's theory also accounts for the behaviour at different wavelengths. Quite simply, the wavelength has to be short enough so that each photon carries sufficient energy to release an electron.

As a historical aside it is interesting to note that Einstein's work on the photoelectric effect appeared in the same year as his special theory of relativity. Surprisingly, it was a further seventeen years before he received recognition from the Nobel committee, and even then the citation only specifically mentioned his work on the photoelectric effect. It seems that the delay was due primarily to the startling nature of his predictions. In particular, a number of prominent scientists refused to believe the theory of relativity. Seventeen years on, the photoelectric effect seemed a much safer bet. Even so, the full consequence of this theory, the interpretation of a light beam as a sequence of particles, was not realized until another two years later.

Louis de Broglie initially entered the Sorbonne to read history. He soon switched to physics, and it was in his doctoral thesis in 1924 that he presented his famous postulate. If light has particle as well as wavelike properties, he reasoned, then should not particles of matter sometimes exhibit wavelike behaviour? What exactly does this mean? How do we conceive of a particle behaving like a wave?

To answer this let us think how we could devise an experiment to observe the wavelike properties of an electron. One possibility, in principle at least, would be to repeat Young's double slit experiment using electrons instead of light. The apparatus could perhaps resemble an ordinary television tube, as shown in Figure 7.3. At one end of the tube is an electron gun which fires out a stream of electrons. At the other end is the screen which is coated with a fluorescent material. When a fast-moving electron strikes one of the atoms on the screen it

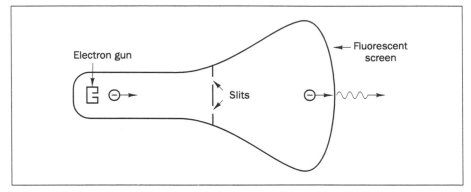

Figure 7.3 An apparatus to demonstrate the wavelike nature of electrons.

produces a small amount of energy which is released as a photon. Part way along the tube we place a barrier in which there are two extremely narrow slits. Now we adjust the electron gun so that just one electron passes through the system at a time. Let us follow one of these electrons as it moves along the tube. If it is to proceed past the barrier it must pass through either one or other of the slits. It then carries on to hit the screen and produce a single photon. We record the position at which the photon is emitted and repeat the experiment many more times. If we now analyse the positions at which the electrons have struck the screen, the results are most surprising. There are distinct areas in which many electrons have been recorded, separated by bands where virtually no electrons have been seen. This is precisely the same result that we would expect to get with light waves, and yet we are using electrons! We explained this behaviour for light waves by saying that the waves must pass through both slits simultaneously. This does not seem to make any sense if we try to apply the same argument to electrons. Surely an electron cannot pass through both of the slits, can it?

We could maybe resolve this problem by suggesting that there are two electrons, one of which passes through each of the slits. They may then interfere with one another to produce the observed pattern. However, we have already ruled out this possibility by requiring that the electrons are sent through the apparatus one at a time. Perhaps the answer lies in trying to determine through which slit each electron passes. The easiest approach would be simply to close off one of the slits. However, if we do this the interference pattern is lost and we obtain just a blurry image of the single slit as we would if we used light. Somehow, although the electron passes through just one of the slits, it must know whether the other one is open or closed. Maybe we need a more subtle approach. We revert to using two slits, but this time we shine a beam of light across the slits. In this way, as the electron passes through the light beam we observe a short flash from the vicinity of the slit through which the electron has passed. This enables us to determine which option the electron has taken, but once again we find that the interference pattern is lost. The problem this time is that the electron must interact with a photon in order for us to see the flash, and in doing so the electron is knocked off course. Can we reduce the energy of the photon so that the disturbance of the electron is negligible? We have seen that the energy of a photon decreases as the wavelength of the light increases. Consequently, using light of a longer wavelength should enable us to recover the interference pattern. The problem is that as the interference pattern begins to reappear the wavelength of the light is now comparable with the spacing between the slits. As a result the flash of light is so blurred that we are unable to determine through which of the slits the electron has passed!

We appear to have encountered a fundamental problem. It arises as a consequence of Heisenberg's uncertainty principle, which states that there is an absolute limit to the precision of the information that we can determine about a system. In this case it means that if we wish to observe the interference pattern generated by the electrons, then we are not allowed to know the route chosen by any particular electron.

We are left with a rather confusing situation. Is an electron really a particle or is it a wave? The answer is that it is neither one nor the other. The two descriptions complement each other. We can get some feeling for what this statement means by imagining how the inhabitants of a two-dimensional world would comprehend a three-dimensional hairbrush. If the brush is placed so that the bristles penetrate their flat land, then the inhabitants would perceive it as a large number of small objects. If we move the brush, then all of the small objects move in unison. To the flatlanders it would appear that there is some mysterious force which maintains a sort of equilibrium distance between each of the objects. However, if we push gently on just one of the bristles, then we can make it move whilst the rest remain stationary. How would the flatlanders interpret these two seemingly contradictory observations? They might come up with a theory explaining how the objects move as a single entity under certain circumstances, and as individual particles in other cases. Of course, from our privileged viewpoint, having access to the third dimension, we can see that neither description in itself is complete. We can regard the wave and particle properties of matter in a similar way. If we could free ourselves from the constraints of our three-dimensional universe, we would be able to perceive these two distinct types of behaviour as a single phenomenon. However, on the whole it is much easier to relate to these two more familiar concepts and choose whichever happens to be the most appropriate description in a given situation. There is, however, one important proviso. Although both the wave and particle aspects of the electron may be observed in the same experiment, they cannot both be seen at the same instant. We discovered this in the Young's slit experiment. The electron behaves like a wave when it passes through the slits, and we therefore obtain an interference pattern. If we instead attempt to see the particle aspect of the electron by determining which slit it passes through, then we lose its wavelike properties.

It is all very well to say that an electron behaves like a wave, but an electron has mass and an electric charge. Does this mean that the mass and charge of an electron are spread out over the extent of the wave? This would be crazy. It would mean that if we isolated just part of the wave, we would obtain a fractional part of an electronic charge. How then do we interpret an electron wave? The answer is that the wave itself does not have any substance. It is a probability wave. When we talk about an electron wave, the amplitude of the wave at a particular point tells us the probability of finding the electron at that point. Thus, when an electron encounters the two slits we can picture part of the probability wave passing through one slit and part through the other. The two halves of the probability wave then recombine to produce regions on the screen where there is a high probability of finding the electron, and other regions where the probability is very low. We cannot predict with certainty where a specific electron will strike the screen, but if we perform the experiment a hundred times we know that the vast majority will end up in the regions of high probability, and this explains the familiar pattern of dark and light bands.

This then is a brief description of one aspect of quantum theory known as wave–particle duality. The ideas are not easy to grasp as they seem to contradict our intuitive understanding of the physical world. It leads us to question the physical models we have already developed. In all the discussions so far we have always assumed that an electron is a small solid object. How will it affect our understanding of these systems if the electron behaves like a wave? In fact, how can we be sure that any of our classical physical models are valid? According to de Broglie's theorem it is not just electrons and other small particles which have a wavelike nature. Dust particles, golf balls and even the planet Earth all have wavelike characteristics! All is not lost, however. Niels Bohr recognized the disparity between quantum theory and classical physics, but also realized that in many systems classical physics gives excellent agreement with experiment. Consequently, he argued that in such cases the predictions of quantum theory must be indistinguishable from those of classical physics. This he called the correspondence principle. We can easily see how this principle operates. In de Broglie's theorem the wavelength of a particle is inversely dependent on the mass of the particle. So, although a golfball does have a wavelike nature, the deviations predicted by quantum theory are of the order of a million billion billion billionth of a metre. To put this in perspective, an atom is slightly less than a billionth of a metre across. We encountered another example of the correspondence principle in the previous chapter. The fact that an electron moves unhindered through a perfect crystal is a prediction of quantum theory. However, because of the thermal movement of the atoms, the electron is frequently deflected from its course. The result of many such interactions is virtually identical to that

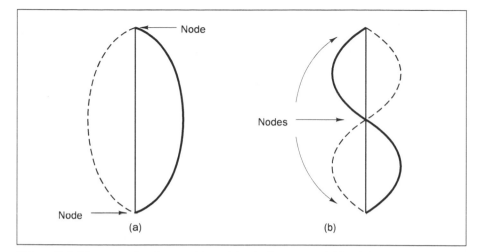

Figure 7.4 Waves confined on a string: (a) a half wave, and (b) a full wave. The nodes are the points which remain fixed at all times.

obtained from the classical picture of a small particle rattling around a rigid lattice of atoms. On the whole, then, classical physics is generally more than adequate for describing the behaviour of relatively large systems. This suggests that if we want to observe the wavelike properties of an electron we must somehow confine it within a small region.

Let us first consider how we can confine a wave using a more familiar example of wave behaviour. Imagine a stretched string which is free to vibrate. A string on a harp is a good example. The string is held firmly in position at both ends and is tensioned so that if we pluck the string approximately midway along its length it produces a particular note. Figure 7.4(a) illustrates the shape of this string as it vibrates. The string oscillates backwards and forwards about its original position, with the maximum deviation occurring midway along the string. The two end points do not vibrate. Such points on a wave are called nodes, and since these nodes always remain at the same position on the string this is called a standing wave. In this case we can both see the wave visually and hear the sound wave that it produces in the air.

To be precise we have actually trapped only half a wave since a whole wave would start at a node, pass through a maximum to another node, and then through a minimum to a third node. We can produce such a wave on the stretched string simply by taking hold of the string approximately a quarter and three-quarters of the way along its length and pulling in opposite directions. From Figure 7.4(b) we can see that the length of this wave is half that of the original wave, and the pitch of the note produced is correspondingly an octave higher. Further tones can be produced in a similar way (given access to a suitable number of hands) so that one and a half or two wavelengths, or indeed any

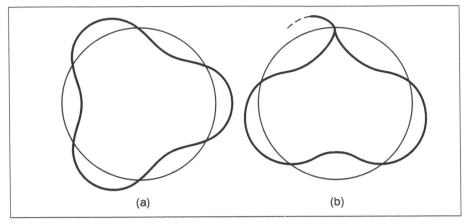

Figure 7.5 (a) An integral number of wavelengths in a loop interfere constructively. (b) A non-integral number interfere destructively.

multiple of half wavelengths, are confined on the string. However, the range of notes that are available from this one string is strictly limited. It is not possible, for example, to produce a note lower in tone than that corresponding to a half wavelength. This is obvious since we can see that the separation of the nodes would then have to be longer than the length of the string. Nor is it possible to produce any notes corresponding to anything between a half and a whole wavelength. So our wavetrap is very selective about the waves that it contains.

Let us now alter the geometry. We take a stiff wire containing a single whole wave and bend the wire back on itself to form a circular loop. What happens to the wave? We can find out by following the wave around the wire, starting at an arbitrary point on the wire. One whole wavelength takes us all the way around the circumference of the wire loop. We are now back at our starting-point, and since we have travelled exactly one wavelength, the wire is displaced by the same amount as when we started. To use the terminology that we introduced in explaining Young's double slit experiment, we can say that the wave interferes constructively with itself, and so the vibrations will continue indefinitely. As Figure 7.5(a) shows, a similar condition is obtained if the wire contains three whole wavelengths, and by applying the same reasoning it is apparent that any integer multiple of whole wavelengths will produce the same results. The question is, will the outcome be the same if we have an odd half wavelength? This requires a bit more thought. Suppose that we have a straight wire containing two and a half wavelengths. This is perfectly admissible since each end corresponds to a node. However, when the wire is bent into a circular loop we find that the oscillations die away rapidly. Why is this? The answer is given visually in Figure 7.5(b). If we again follow the wave around the loop, when we arrive back at the starting-point the displacement of the wave is now different from that of the original wave. In fact, we can see that if the original wave was displaced in one direction, then the wave which has travelled once around the loop is found to be displaced by precisely the same amount but in the opposite direction. As a result the two waves interfere destructively and cancel each other out. We therefore conclude that the only waves which can be contained in the wire loop are those for which the circumference of the loop is equal to an integral number of wavelengths.

What is the connection between these vibrating wires and the wave nature of an electron? Let us begin by considering an atom. After all, it is hard to imagine an electron being confined to any region smaller than an atomic orbit, and so we expect the wavelike nature of the electron to be observable in this context. In Chapter 1 we stated that electrons are only allowed to occupy certain orbits corresponding to specific allowed energies. We are now in a position to be able to offer some justification for this rather ad hoc proposition. The de Broglie wavelength of an electron in the smallest orbit turns out to be exactly equal to the circumference of the orbit predicted by Bohr. Similarly, the second and third orbits are found to contain exactly two and three de Broglie wavelengths, respectively. From this picture it now becomes clear why only certain orbits are allowed. In direct analogy with the wire loop, the electron orbit must contain an integral number of de Broglie wavelengths. If it did not then the wave would interfere destructively with itself and cancel itself out! In other words, the probability of finding an electron in any other orbit is zero.

Obviously, the wave nature of an electron plays an important role in an atom, but the question we are most interested in is: can we create an artificial structure which exploits this property? The wavelength of an electron in a semiconductor is typically about ten nanometres (a nanometre is one millionth of a millimetre). This corresponds to a region about fifty atoms across. Consequently, if we confine the electron to a layer of this thickness we should be able to observe quite strong quantum effects. We can produce such a structure using the gas deposition (CVD) method discussed in the previous chapter. We begin by growing an aluminium gallium arsenide (algas) crystal. The supply containing the aluminium gas is then shut of so that a layer of pure gallium arsenide forms on the surface. Once we have the required thickness of gallium arsenide the aluminium supply is turned back on and we resume growing aluminium gallium arsenide. Figure 7.6 illustrates schematically the effect of this structure on the energies of the conduction electrons. As we discussed in Chapter 6, the conduction electrons will accumulate in the gallium arsenide layer since once they fall into these lower energy levels they are effectively trapped there. It is rather like dropping a ball into a well. It is easy to put it in there, but very difficult to get it back out. Using this analogy we say that the electron is in a quantum well. However, the term 'well' is a bit misleading since the electron is actually only trapped in one direction. It is still free to move in the directions parallel

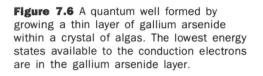

Figure 7.6 A quantum well formed by growing a thin layer of gallium arsenide within a crystal of algas. The lowest energy states available to the conduction electrons are in the gallium arsenide layer.

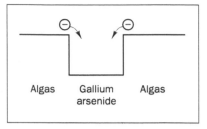

to the interface layers, just like the electrons in the modulation doped transistor. A better description is to think of the electron as being confined in a steep-sided gorge, like the Grand Canyon. We can picture a current of electrons moving along this channel, just like the River Colorado flowing through the Grand Canyon. However, we must be very careful when using analogies to describe quantum phenomena. It goes without saying that we expect the River Colorado to flow along the very bottom of the Grand Canyon. We would be surprised, to say the least, to find it flowing through mid-air a hundred metres above the valley floor. And yet this is precisely how the electrons appear to behave.

The problem is that we are still thinking about electrons as being small discrete particles, when really we should be considering electron waves analogous to the waves on a stretched string. We have seen that the longest wave that can be accommodated on such a string is one whose wavelength is twice the length of the string. Consequently, if the wavelength of an electron at the bottom of the well is more than twice the width of the well, then there is no room for such an electron within this region. However, we know from the discussion about the photon that the wavelength of a wave decreases as its energy increases. We therefore find that an electron slightly higher up the well has a wavelength exactly twice the size of the well. As we can see from Figure 7.7, such an electron wave has a node at each side of the well and therefore fits perfectly.

Although this is the lowest energy state that an electron is allowed to occupy, by analogy with the stretched string we can see that certain higher energy states are allowed for which the well width corresponds to an integral number of half wavelengths. We conclude that electrons in a quantum well are only allowed to occupy certain discrete energy levels, a result which is reminiscent of the energy states in an atom. However, there is a significant difference between the two systems. The energy levels in an atom are determined by nature. Consequently, we find that all atoms of a given type have precisely the same set of energy levels. We mentioned this in Chapter 1, giving the example of a sodium street lamp which emits a characteristic yellow light. In contrast, the energy levels in a quantum well are governed by the dimensions of the well. We can easily demonstrate this by considering what happens if we manufacture a quantum well which is somewhat narrower than the one considered in the above example. The

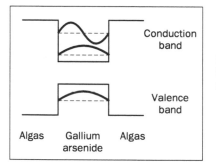

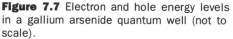

Figure 7.7 Electron and hole energy levels in a gallium arsenide quantum well (not to scale).

longest wavelength of an electron allowed in the well is now shorter, and correspondingly the energy of this state is higher. It is not so obvious, but the energies of the discrete states are also dependent on the depth of the well. If we increase the aluminium content of the algas layer, then the well becomes deeper, and this also raises the energy of the lowest state. By the same reasoning, we find that a wider, or shallower, well can accommodate lower energy electrons. This means that we can actually choose the energies of the electrons which are allowed in the well. By making small changes during the growth of such a structure we are actually changing the quantum states. This is engineering on a quantum scale! The implications of this statement are hard to appreciate at this stage, but we will encounter several examples of the applications of this effect during the subsequent chapters.

It is not just the conduction electrons which are affected in this way. In creating the gallium arsenide–algas interface we found that higher energy levels are available to holes in the gallium arsenide than to holes in the algas. This means that the holes also accumulate in the gallium arsenide layer, and similarly exhibit wavelike behaviour (see Figure 7.7). In this way the energies available to electrons near the bottom of the conduction band and to holes near the top of the valence band are controlled by the quantum well. Since many of the properties of semiconductors are particularly dependent on these states, we might ask what effect this has on the material. One of the most noticeable properties to be affected is its optical behaviour, and accordingly one application of the quantum well is in a laser.

How does a quantum well laser differ from the ordinary diode laser described in Chapter 3? Firstly, in the diode laser, recombination of conduction electrons and holes takes place across the whole of the depletion layer. This is a very wasteful process because it is only possible to achieve a population inversion over a small part of this region. Consequently, a large proportion of the recombination occurs by spontaneous emission rather than stimulated emission. (If you need to remind yourself of the meaning of these terms, refer either to Chapter 3 or to the Glossary.) In comparison, the electrons and holes in a quantum well are concentrated in a narrow layer, and virtually all of the recombination takes place within this region. This means that the quantum well laser is far more efficient than a diode laser. In fact, the same argument applies even if the central gallium arsenide region is too wide for quantum effects to be observable: it is only the presence of the abrupt confining barriers which is important. Since each barrier is created by a junction between two different materials, it is referred to as a heterojunction. Accordingly, structures of this form which do not exhibit significant quantum effects are called double heterostructure lasers.

Other important properties of the quantum well laser arise as a direct result of the quantization of the energy levels in the well. One such feature is the ability to 'tune' the wavelength of the emitted light. We know that the wavelength of this light corresponds to the difference in energy between an electron at the bottom of the conduction band and a hole at the top of the valence band. In a conventional gallium arsenide laser the band gap is a fixed quantity, and so the light which is produced always has exactly the same wavelength. However, in a quantum well the energy is determined by the position of the lowest allowed electron state and the highest allowed hole state, as shown in Figure 7.7. Since these energy levels can be altered by varying the width and depth of the well, the wavelength of the emitted light changes accordingly.

Here we have encountered the first example of quantum engineering, or band gap engineering as it is more commonly called. The achievement is so amazing that it is worth reiterating the above statement in order to emphasize what has been achieved. By varying the dimensions of the well we are actually manipulating the quantum states within the well. Perhaps even more surprising is the fact that such subtle changes produce an effect which can be observed without sophisticated equipment. For example, if we construct the quantum well system using the alloy gallium arsenide phosphide, which emits light in the visible spectrum, then the change in energy of the band gap is apparent from the colour of the light that the device emits.

There is a further way in which quantum confinement of the electrons and holes affects the performance of the laser. To understand this behaviour it is helpful to think of the grains of sand in an egg timer. When we start the egg timer, all the grains of sand are contained in the top half of the vessel, so we have an instance of population inversion, just as we require for a laser. Of course, not all of the grains can instantaneously fall into the lower half of the vessel. (It wouldn't be much use as a timer if they did!) Instead they are restricted by a narrow passage so that only a few grains can pass through at a time. Those grains higher up the pile have to wait their turn. A similar situation exists for the conduction electrons in a crystal of gallium arsenide. In stimulated emission a photon will encourage a conduction electron to recombine with a hole in the valence band if by doing so it produces another photon of the same energy. This is the case only if the electron is at the very bottom of the conduction band and a hole exists at the very top of the valence band. However, the electron states are not distributed uniformly throughout the energy bands. The density of states is highest near the middle of the band and very low near the upper and lower extremes of the band. In other words there are only a handful of states at the lowest energies in the conduction band, and so only a small number of conduction electrons can occupy these states at any one time. Electrons at higher energies have to wait for these states to become vacant before they can produce a photon of the desired energy.

How does the presence of a quantum well affect this picture? We have represented the density of states at the bottom of the conduction band of a bulk material by an inverted cone, but in a quantum well there are no allowed electron states at the very lowest energies. In the analogy with the egg timer we can represent this by simply slicing the tip off the cone. We are now left with a gaping hole, and the sand pours through with ease. What this really means is that the lowest conduction state in a quantum well is broad so that many electrons can be accommodated in this state at once. Similarly, the top of the valence band has a capacity for a large number of holes, which all have the same energy. This means that it is possible for many electrons to recombine simultaneously with holes in the valence band and produce photons with identical energies. These two scenarios are shown schematically in Figure 7.8.

From the above discussion we can see that the effect of confining the electrons and holes in a quantum well greatly enhances the probability of stimulated emission occurring. Consequently the efficiency is improved, with the result that much smaller currents are required than in a standard diode laser. Typically, the current density, in other words the rate at which electrons pass through a given area of the device, is reduced by a factor of several hundred. This in turn means that far less heat is generated by a quantum well laser, opening up a whole host of opportunities for the use of such a device. Laser printers and compact disc players already make use of this technology. By taking full advantage of the small physical size of these devices it is possible to integrate quantum well lasers into conventional logic chips, and even to create specialized chips containing literally millions of tiny lasers. Such advances pave the way for an entirely new technology which we will examine in more detail in Chapter 11.

✳

We have seen that the electrons in a quantum well are confined in one direction but are still free to move along the directions parallel to the interfaces. In this way we can think of the electrons as being constrained to a surface. They have only two dimensions in which they can move, compared to the three dimensions available to electrons in a bulk material. The same reduced dimensionality applies to electrons confined at the interface in a modulation-doped FET, and even in

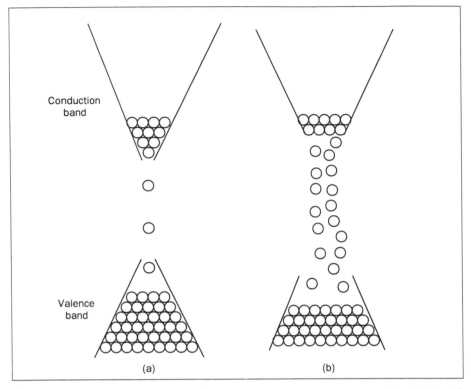

Figure 7.8 A schematic representation of recombination showing how the number of states available to electrons at the extremes of the bands affects the efficiency of this process (a) in a bulk material, and (b) in a quantum well.

some cases to electrons confined in the narrow channel of a conventional MOSFET. On the whole this restriction does not play a vital role in our understanding of the operation of the MOSFET, but it was while investigating such a structure in 1980 that Klaus von Klitzing, Gerhard Dorda and Mike Pepper uncovered one of the most astounding and unexpected discoveries of recent years: the quantum Hall effect. The following discussion of the quantum Hall effect is quite lengthy. For any reader who has already progressed this far through the book there is nothing which is conceptually difficult. However, those who are faint at heart or get lost on the way may skip to the last paragraph in this chapter without significantly handicapping themselves in the following chapters.

Before we can ask what the quantum Hall effect is, we must first understand something about the ordinary Hall effect. To do this we need to examine how the electrons in a solid are affected by the presence of a magnet. Suppose that we take a crystal of semiconductor doped with donors so that it contains a majority of conduction electrons. If we place this sample in a strong magnetic field so that the poles of the magnet are above and below the crystal, then we find that the free electrons move in horizontal circular orbits. They do this because the effect of the magnetic field is to push an electron sideways, the force being always at right-angles to the direction in which the electrons are currently travelling. In contrast, we know that the effect of applying an electric field is to tilt the energy bands so that the electrons tend to move down the slope. What then is the effect of applying both a magnetic field and an electric field simultaneously? To answer this question let us suppose that this book represents the semiconductor crystal. Begin by holding the book horizontally in front of you. If we apply a voltage so that the top of the page is negatively charged and the bottom is positively charged, then this has the effect of tipping the book slightly towards you so that the electrons flow down the page. If we now place a magnet so that the North pole is above the book and the South pole is below it, then the magnetic field causes the electrons to be deflected towards the left-hand side of the page.

There is also another force at work here. As the electrons accumulate on one side, a negative charge builds up which acts to repel other electrons from entering this region. The argument is similar to the one used in Chapter 3 to determine the redistribution of electrons at a p-n junction. Each electron which moves over to the left-hand side of the sample makes it harder for subsequent electrons to follow. A balance point is achieved when the repulsive force is equal to the magnetic force pushing the electrons into this region. Since the equilibrium condition corresponds to an excess of negative charge on the left-hand side, a voltage difference exists between the two sides of the sample. This is the Hall voltage.

Using this explanation it is straightforward to determine how the strength of the magnetic field affects the Hall voltage. If the magnetic field strength is increased, then more electrons can be pushed on to the left-hand side of the sample before the balance point is reached. Accordingly, the voltage difference between the sides of the sample increases. Conversely, a decrease in the magnetic field strength results in a reduction in the Hall voltage. We therefore find that the Hall voltage increases in direct proportion to the magnetic field strength.

An alternative description, and one which we will use when discussing the quantum Hall effect, is to consider the resistance of the sample. To use the analogy of the book again, the electrical resistance is normally defined as being equal to the voltage difference between the top and bottom of the page divided by the electron current flowing down the page. In a similar way, we can define the Hall resistance as being the voltage difference between the two sides of the sample divided by the current flowing from top to bottom. (There is of course no current flowing across the sample because the electrons cannot escape at either side.)

This describes the Hall effect, which was first noticed by Edwin Hall in 1879. It was many years before the results were fully understood. One of the most puzzling features at the time was the observation that in some samples the polarity of the Hall voltage seemed to be the wrong way round. Using the same configuration as described above, it was found that the left-hand side of the sample became positively charged in these cases, as though there were positive carriers present in the material. Of course, we now know that these positive carriers are what we call holes, but these particles were unknown until the 1920s. In more recent years the Hall effect has proved to be a very useful tool, particularly in the study of semiconductors. It not only allows us to determine whether the majority carriers are positively or negatively charged, but by measuring the magnitude of the Hall resistance it is possible to determine the number of carriers in the material. Given its routine use in research and industry, it is all the more surprising that something unexpected should be observed a hundred years later. The reason why it took so long to discover the quantum Hall effect is that it is only recently that it has become possible to produce the required conditions.

There are three factors essential for the observation of the quantum Hall effect. Firstly, and most importantly, the movement of the electrons must be restricted to only two dimensions. We have seen that this can be achieved by ensuring that the width of the confining region in the third dimension is comparable with the wavelength of the electrons. To visualize this we can still think of the book as representing our sample of semiconductor, but now the electrons are constrained to the surface of one page. The second and third criteria are for a very intense magnetic field and for the sample to be maintained at an extremely low temperature, within a degree or two of the lowest possible temperature, absolute zero.

What is so special about the properties of the material under these conditions? We have seen that if we apply a magnetic field (but no electric field) to a crystal containing free electrons, then these electrons move in circular orbits. In fact, the radius of the orbit depends on the energy of the electron. In the ordinary case there is virtually no restriction on the energy of a conduction electron, and so all sizes of orbit are permissible. However, if we confine an electron in a quantum well (or some other structure of similar dimensions), then only certain discrete energy levels are allowed. Consequently, only certain sizes of orbit are permitted for an electron in a quantum well subject to a magnetic field. In this context the corresponding energies are called Landau levels.

It is tempting to make a comparison with the electron orbits in an atom, but to do so would be misleading since there are some significant differences. For instance, the number of electrons which can be accommodated in each Landau level is dependent on the strength of the magnetic field. This number is also dependent on the ratio of two fundamental constants, the charge on an electron

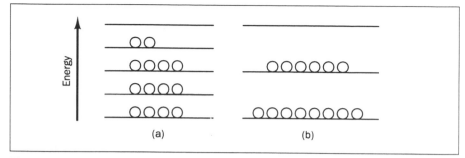

Figure 7.9 The occupancy of electrons in Landau levels. In (b) the magnetic field strength is twice as large as that in (a).

and Planck's constant. (Planck's constant was first introduced by Max Planck—it relates the energy of a photon to its wavelength, and appears in virtually every equation in quantum theory.) It is significant, however, that the number is independent of the type or purity of the material. We will discover the relevance of this statement shortly.

The magnetic field strength also affects the Landau levels in another respect: it determines the energy separation between adjacent levels. The effect of doubling the intensity of the magnetic field is illustrated schematically in Figure 7.9. The separation between the energy levels increases by a factor of two, but twice as many electrons are now allowed in each level. (In a real system the number of electrons per level is typically of the order of a million billion.)

To appreciate why the large magnetic field and low temperature are required we need to think back to the intrinsic semiconductor that we discussed in Chapter 2. In the idealized case the valence band of a semiconductor is full, while the conduction band is empty. However, we found that in a crystal of pure silicon at room temperature the thermal energy is sufficient to excite some electrons into the conduction band. Similarly, we expect to find a number of thermally excited electrons in some of the higher Landau levels even when there are spaces in lower levels. However, in order to observe the quantum Hall effect it is crucial that the electrons occupy the lowest possible energy states. This requires extreme conditions because the energy separation between adjacent Landau levels is very small. Even in an intense magnetic field the energy differences are typically a thousand times smaller than the band gap of silicon. This means that a great effort has to be made to reduce the thermal energy of the electrons, hence the extremely low temperatures.

Having described the conditions required, let us now see what happens under these circumstances. There are two surprising results which constitute the quantum Hall effect. First of all we will examine the resistance of the sample, but before we do this we should briefly remind ourselves of the causes of resistance. In the previous chapter we found that resistance occurs because an electron is deflected by the movement of the atoms and by impurities in the sample. Since we are dealing with exceptionally low temperatures, we can assume that the dominant effect is due to the impurities. We also showed that the probability of an electron being deflected is dependent on the number of different states into which the electron can be scattered. Consequently, the resistance of the sample is directly proportional to the number of vacant states available.

Now, let us suppose that the intensity of the magnetic field is such that only two of the Landau levels are occupied, as shown in Figure 7.9(b). The lowest level is filled, whilst the second level is partially filled. When an electron in the second level interacts with an impurity it is scattered into a new state. We know that this new state must be one of the vacant states within the same level, since the conditions that we have imposed ensure that the electron cannot gain enough energy to move up into the third Landau level. What happens if we now decrease the magnetic field strength slightly? This has the effect of reducing the number of electrons allowed in each Landau level. In consequence, the lowest level is overpopulated, so some are forced to move up to occupy vacant states in the second level. Since there are now fewer vacant states into which the electrons can scatter, the resistance of the sample decreases. As the magnetic field strength is reduced further we reach a situation where the second level becomes precisely filled. There are now no vacant states, so it is no longer possible for an electron to be scattered into a different state. We therefore conclude that the sample offers no resistance to the flow of electrons. In practice the resistance may not be precisely zero, but it is considerably lower than the best metal conductors at the same temperature. (Remember, we are talking about a semiconductor!)

The second phenomenon associated with the quantum Hall effect is still more astounding. So far we have considered the resistance along the length of the sample. What about the Hall resistance between the two sides? Let us consider the same scenario as above. The Hall resistance decreases as the magnetic field strength is reduced (as we would expect for the ordinary Hall effect), until we reach the point where the two lowest Landau levels are filled. What happens if we now reduce the magnetic field strength a bit further? The two lowest levels can no longer accommodate all the electrons, so we would expect a few to spill over into the third level. However, this is not quite correct. We stated previously that all of the electrons within a Landau level have precisely the same energy, but this does not take account of the presence of impurities in the sample. For example, let us consider the effect of a positively charged donor ion. A nearby electron will be attracted towards this ion, and as a result will have a slightly lower energy than the other electrons in the Landau level. The picture is therefore slightly more complicated than we first assumed. In addition to the Landau levels there will be a few intermediate states associated with the presence of impurities. This is shown schematically in Figure 7.10. (You may notice that the picture

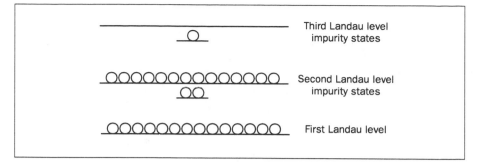

Figure 7.10 The presence of impurities gives rise to additional states in between the Landau levels.

is somewhat reminiscent of the appearance of allowed states in the band gap of a semiconductor due to doping—see Figure 2.9.)

Armed with this new knowledge, let us now return to the problem that we were considering before. What happens if we decrease the magnetic field strength slightly below the point at which the two lowest Landau levels are filled? The two lowest states can no longer accommodate all of the electrons, and so a handful are forced to move to higher energy levels. The next available states are the intermediate states, and so this is where the electrons which are evicted from the second level end up. The importance of this result is that an electron in one of these intermediate states is constrained to the vicinity of the impurity. We describe this by saying that the electron is in a localized state. This means that it cannot be deflected by the magnetic field, and so does not affect the Hall resistance. Consequently, if we continue to decrease the magnetic field strength, the Hall resistance remains constant until all of these localized states are filled. If we examine how the Hall resistance varies with the intensity of the magnetic field, we therefore find that there are 'plateaus' each time a Landau level is filled. A typical example is shown in Figure 7.11.

This in itself constitutes an interesting effect, but what is most surprising is the degree of accuracy with which these results can be repeated. In general, measurements of resistance or of the Hall resistance are particularly sensitive to many factors such as the temperature, the impurity content and even the shape of the sample. However, the plateau resistances of the quantum Hall effect are virtually identical for different samples and even different types of material. It turns out that the Hall resistance at these plateaus depends only on the two fundamental constants that we mentioned earlier, the electron charge and Planck's constant, and on the number of completely filled Landau levels. An application has already been found for this effect in the field of metrology, the branch of physics which is concerned with the accurate measurement of fundamental quantities. Since the amount of charge on an electron and Planck's constant are already known to an accuracy greater than one part in a million,

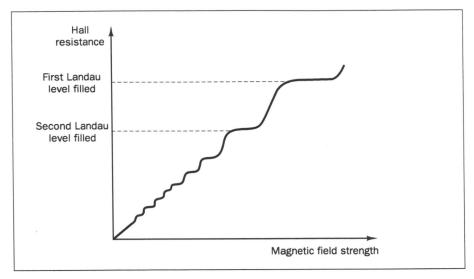

Figure 7.11 The quantum Hall effect is characterized by distinct 'plateaus' in the Hall resistance corresponding to the filling of each Landau level.

the quantum Hall effect provides a highly accurate standard from which other resistances can be measured.

The discovery of the quantum Hall effect has been the source of great excitement in the physics community because of the deep insight that it sheds on the quantum world. In fact, the quantum Hall effect appears to be just the tip of the iceberg. Within two years of its discovery a still more bizarre phenomenon was discovered: the fractional quantum Hall effect. The observations appear to be similar to the integer quantum Hall effect, with plateaus occurring in the Hall resistance and zeros in the resistance, yet these occur when the Landau levels are fractionally filled. The experimental requirements are even more extreme, with temperatures being reduced to a tiny fraction of a degree above absolute zero. However, the physics describing these effects is remarkably different and would take us well beyond the scope of this book.

We have seen that by restricting electrons to just two dimensions the properties of a material are altered dramatically. Can we reduce the freedom of the electrons still further? Yes, in principle: we could grow a thin column of gallium arsenide on to an algas surface and then surround it with more algas. In this way we form a quantum wire. An electron is free to move along the axis of the wire, but exhibits wavelike properties in the other two dimensions. Producing such a structure in practice is rather more difficult, but many ingenious methods are under development. The natural culmination of this process is a quantum dot, a tiny box of material containing low-energy electrons amidst a matrix of higher-energy material. In such a structure the electron is confined in every direction. It has zero dimensions of freedom. What is the point of such a structure? If an electron is confined in a quantum dot, will it remain there forever? No, the electrons can get out, but to find out how we need to examine another aspect of quantum theory.

8

The Joy of Tunnelling
From Superatoms to Superlattices

L ET us begin by considering the following thought experiment. Suppose we take two identical buckets. We fill one with water and leave the other one empty. They are placed side by side and left undisturbed for a short while. Will the water levels have changed when we return? Of course not? We might even wonder why someone should ask such a stupid question. However, suppose we consider a seemingly analogous case in which the buckets are replaced with quantum wells. A number of electrons are initially placed in just one of the wells. Surprisingly, we find that after a short while the numbers of electrons in each well are equal. How does this happen?

Of course, a small proportion of the electrons will gain enough thermal energy to allow them to jump out of the well, in a way similar to that in which an electron in a semiconductor is occasionally promoted into the conduction band. A number of these electrons may then find their way into the adjacent well. This process can therefore explain the appearance of a few electrons in the second well, but it cannot explain how half of the electrons from the first well suddenly appear in the neighbouring well. There must be some other mechanism at work. It seems that the only way in which we can explain this behaviour is if the electrons somehow tunnel through the intervening barrier that separates one well from the other.

How can we understand this process of an electron tunnelling through a seemingly impenetrable barrier? Let us go back and examine the model of a standing wave on a string. We said that the lowest mode of vibration is one in which half of a wavelength fits exactly onto the string, so that the nodes of the wave are at the two end points of the string. Now let us try to imagine a rather unusual room in which the walls are made from a jelly-like material. We stretch the string across this room and somehow fasten the ends to two opposing jelly walls. If we now pluck the string at its centre the appearance of the vibrating string is similar to before, but since we no longer have rigid anchoring points the ends of the string also move slightly. The wave must, therefore, extend some short distance into the jelly walls. If the walls are thin, then it is possible that the wave actually extends right the way through the wall into the adjoining room.

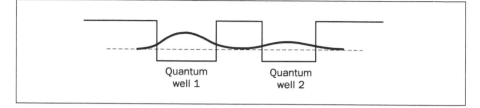

Figure 8.1 Since an electron does not terminate abruptly at the edge of a quantum well, the wave associated with an electron in well 1 may leak into well 2.

This suggests that if we position another stretched string between the walls of this second room, it too will begin to vibrate in sympathy with the first string.

We can use a similar picture to visualize how an electron tunnels through a barrier. In the previous chapter we described how a conduction electron can appear to be trapped in a very narrow layer of gallium arsenide sandwiched between two much thicker layers of algas. Using the analogy with the stretched string we said that the nodes of the electron wave must occur precisely at the edge of the quantum well. However, it turns out that the above example is closer to the truth. The electron wave does not terminate abruptly at the edge of the quantum well, but actually leaks out of the well slightly. This is a most surprising result. We should remind ourselves that the existence of the quantum well is due to the lower energy electron states available in the gallium arsenide layer. There are no states of similar energy available in the algas layer since this corresponds to the forbidden energy gap, and yet if the electron wave exists in this region there must be some probability for the electron actually being in this forbidden zone!

Let us now take the argument a stage further. If we make one of the algas layers very thin and construct another gallium arsenide quantum well on the other side, then we obtain the structure in Figure 8.1. We can now see that if we modify Figure 7.7 to allow the electron wave to leak out of the well, then a small proportion of the electron wave may appear in the second well. As the majority of the probability wave is in the original well, it is still most likely that the electron is in this well. However, since both wells are identical, there is no reason why the electron should choose one in preference to the other. This means that an electron which is initially in the first well may suddenly appear in the second well.

This shows how we can understand tunnelling by using a wavelike picture of an electron. An alternative view can be obtained by considering an electron as a particle. To do this we need to employ Heisenberg's uncertainty principle. We encountered this principle previously when examining Young's double slit experiment with electrons. When we attempted to determine the position of the electron, in order to find out which of the slits it passed through, we disturbed the subsequent motion of the electron and so destroyed the interference pattern. This restriction, forbidding the simultaneous measurement of both the position and momentum of a particle, is the most familiar aspect of the uncertainty principle. However, there are several other important consequences. For example, it is found that a particle is allowed to 'borrow' a small amount of energy for a short time, so long as the loan is repaid in full at the end of that time. Let us see how this system works.

Suppose we discover a bank with rather sloppy accounting procedures, which allows customers to be overdrawn without penalty so long as their accounts are in credit at three o'clock every afternoon. We go into the bank in the morning and withdraw £1000 that we haven't got. What shall we do with this free loan? As we are feeling lucky we decide to go to the casino. By midday we are £3000 up, and decide to stay for just one more win. However, Lady Luck leaves us at this point. By half past two we have just the initial £1000 left, so we rush back to the bank before the bank manager notices that the money is missing. It all seems a bit of a waste of time. Surprisingly enough, electrons are constantly performing such futile acts. However, occasionally the results are quite different, as they would have been if we had quit when we were ahead.

If we apply this to an electron in a quantum well, then the electron can borrow the energy that it needs in order to move up to the lowest conduction state in the algas layer. The energy loan lasts for only a million billionth of a second, but this may be just long enough for an electron to find its way into a neighbouring quantum well. The energy loan is repaid, but now the electron is in a different well from the one it started out in. Since classically such a process could not take place, it appears as though the electron has tunnelled through the algas layer.

*

We have now described the process of tunnelling. Three very different applications of this effect will be considered in Chapters 9 and 10. However, for the remainder of this chapter we will examine how tunnelling can be employed to change the physical properties of a semiconductor.

There is no reason why we need to restrict ourselves to just two quantum wells. We can continue to grow alternate thin layers of gallium arsenide and algas as though we are constructing a multilayer gateau with alternate layers of cake and cream. What happens to the electrons in this case? Let us again use the analogy of waves on a string. In this instance we will use people to generate the waves in each well and the string will actually need to be a long piece of rope laid in a straight line along the ground. The participants are spaced at regular intervals along the length of the rope. Each one picks up the rope using two hands and waves their arms up and down so that waves travel outwards in each direction

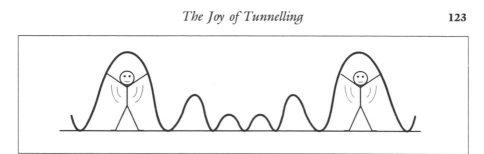

Figure 8.2 When the wave-makers stand close together the waves generated by one person interfere with those from their neighbours.

along the rope. If the participants are initially some distance apart, say twenty metres or so, then parts of the rope will remain in contact with the ground at all times. In other words, the waves will die out before they reach these points. The rope is now laid down on the ground and the participants move closer together before repeating their actions. This time we find that ripples from the waves generated by one person meet similar ripples, travelling in the opposite direction, produced by the next person along the line. Such a situation is shown in Figure 8.2. As the people move still closer together, the waves that they generate interact more and more strongly.

We can visualize the electron waves in a system of quantum wells behaving in a similar way. When the barrier layers between the wells are thick, the electrons in one well are not influenced by those in adjacent wells, and so we effectively have a system of isolated quantum wells. If the barrier layers are made thinner, then the electrons in neighbouring wells interact. What is the result of this interaction? We have, in fact, already considered a similar case back in Chapter 2. We found there that when many atoms are placed close together, the discrete energy levels in the individual atoms interact to produce continuous bands of allowed energies. Although we had not at that stage introduced the concept of an electron having a wavelike nature, the formation of energy bands can be attributed to the interactions between the electron waves. The discrete states in a single quantum well are in some respects similar to those in an isolated atom. However, since we have the freedom to control the energy levels in a quantum well, we can think of it as a 'superatom'. If we place a number of superatoms in close proximity we obtain an artificial crystal or 'superlattice'. Each discrete state broadens into a narrow band of energies called a miniband, as shown in Figure 8.3. Since our superatoms are typically tens of atoms apart, the interactions are much weaker than those between the atoms in a crystal, and so the range of allowed energies in each miniband is only about a hundredth of that in the valence or conduction bands. As in the quantum well, it is only the electron and hole energies near the band gap which are affected, but it is these states which are most important in determining many of the properties of the crystals. Furthermore, we can again tune the properties of the material. We

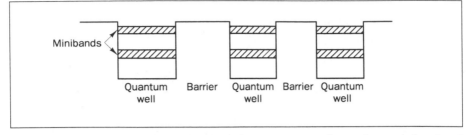

Figure 8.3 Interactions between neighbouring quantum wells produce minibands of allowed energies in a superlattice.

know, for instance, that changing the width and depth of the well alters the allowed energies in the well, and we can also see that by varying the width of the barrier layer separating the wells, we can affect the degree of interaction between the electrons in neighbouring wells. This means that by carefully controlling the atomic layers in a crystal we can produce materials which have properties quite different from those of any naturally occurring materials. This is a most exciting prospect. In the past, designers of microelectronic devices have had a small number of materials from which to choose. In the near future it may be possible for device designers to specify the characteristics they require, and then to manufacture an appropriate material.

Let us look at an example of how a superlattice can be used to produce a new material with novel characteristics. We have seen that semiconductors can emit light as a result of conduction electrons recombining with holes in the valence band. The efficiency with which this process is performed varies considerably from one material to another, so much so that it is usual to divide semiconductors into two quite distinct categories. Direct gap materials produce light efficiently, and so are suitable for use in light-emitting diodes and lasers, whilst indirect semiconductors are very inefficient and unsuitable for such uses. What is it that is so different about these two types of materials?

Each electron in a crystal can be characterized by two parameters, its energy and its momentum. The momentum of a macroscopic object is a fairly familiar concept. It is equal to the mass of the object multiplied by its velocity. If we consider the collisions of billiard balls on a frictionless surface then both the energy and momentum must be conserved in each collision. The same consideration applies when a conduction electron recombines with a hole, although the momentum of an electron in a crystal represents a slightly different quantity: it is related to the inverse of the electron's wavelength. When an electron and a hole recombine, the excess energy is carried away by a photon, which balances the books so far as the energy is concerned. However, a photon of visible light has a much longer wavelength than an electron in a crystal, and so its momentum is negligible in comparison. This means that the emission of a photon does not make any useful contribution to the momentum equation. We arrive at the conclusion that efficient recombination can only take place if a conduction electron has the same momentum as a hole in the valence band. The problem is that we cannot alter the momentum of the electron or the hole—each is determined by the crystal. In direct gap materials, such as gallium arsenide, it so happens that the lowest energy electrons in the conduction band have the same momentum as the holes at the top of the valence band. The situation is quite different in indirect materials, such as silicon. In this case we find that the electrons in the conduction band have different momentum from the holes in the valence band.

The fact that silicon is an indirect material is a considerable drawback. Although silicon technology is the most advanced in terms of integrated circuits, the indirect band gap of silicon severely restricts its use in the rapidly growing field of optoelectronics, which aims to make optimum use of both the electronic and optical properties of semiconductors. The only other semiconductor which is chemically compatible with silicon is germanium, but germanium also has an indirect band gap, as do all alloys of silicon and germanium. This is where the construction of a superlattice helps. If a crystal is constructed by alternating layers of silicon and germanium, with each layer only a few atoms thick, then amazingly enough we can produce a material which behaves like a direct gap semiconductor. Why is there this difference? After all, the superlattice and the alloy differ only in the fact that the atoms in the superlattice are ordered in a particular way, whereas the silicon and germanium atoms are distributed at random throughout the alloy. The reason is that the conduction and valence bands in the alloy crystal are very similar to those of silicon and germanium, and so it is hardly surprising that the alloy behaves like its two constituents. In the superlattice, the electrons and holes which recombine are contained in the minibands. Since these states are unlike any found in bulk silicon or germanium, it is not surprising that the superlattice behaves in a quite different manner.

Although we can predict useful properties for superlattices composed of silicon and germanium, the practicalities of producing such a system are far from straightforward. In particular, there is a problem with the sizes of the atoms. The separation between the atoms in a germanium crystal is about four percent larger than that in a silicon crystal. How does this difference in size affect the growth of a layer of germanium atoms onto a silicon surface? We have briefly mentioned a similar arrangement in Chapter 6 where we found that the growth of indium gallium arsenide on a gallium arsenide crystal produced an internal strain in the crystal. Since such strained layer structures are becoming increasingly important in a variety of applications, it is worth taking a more detailed look at the growth of these structures.

Let us first construct a model of a silicon crystal in which each atom is represented by a standard-sized tennis ball. We begin by arranging the balls in

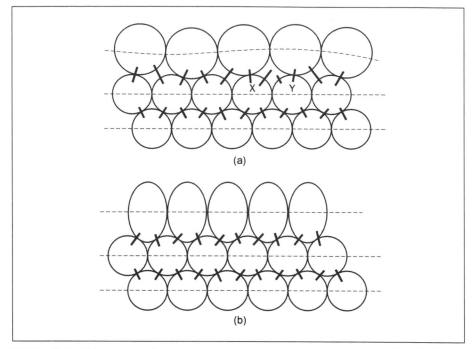

Figure 8.4 When a layer of balls (atoms) are placed on a surface consisting of slightly smaller balls, the balls may either (a) retain their shape, in which case they do not form a plane, or (b) distort their shape in order to maintain the existing structure. The bonds between the atoms are indicated by thick lines. Note that in case (a) there are two atoms (X and Y) which have an unsatisfied bond.

orderly rows across the bottom of a glass tank. We will assume for convenience that a row of one hundred balls can just be accommodated in the tank. Once we have enough balls to cover the bottom of the tank, the first plane of atoms is complete. We refer to this as a plane because the centres of the balls are all at the same level, but of course the surface of this layer is not flat. Consequently, when we form the second plane, the new balls sit in the dimples corresponding to the spaces between the balls in the first plane. As we continue to add more balls, they form successive regularly spaced planes, typical of the arrangement of atoms in a crystal. What happens if we now try to construct a further layer using balls which are four percent larger in diameter? Naively, we might expect to get a similar plane of atoms, but with each row containing only ninety-six balls instead of a hundred. However, a little thought shows that this will not work. Suppose we place one of the larger balls in a dimple on the surface. We would like to place another ball in the next dimple, but these new balls are too large, so the second ball will have to be placed slightly off centre. Since each ball is slightly larger than those in the layer below, as we add more balls they will occupy sites progressively further from their expected positions. Figure 8.4(a) illustrates what happens in a grossly exaggerated case. It is easy to see that the centres of these balls no longer lie in a flat plane.

There is a more serious consequence if we suppose that the atoms of germanium arrange themselves in a similar manner on top of a silicon surface. We know that in a silicon (or a germanium) crystal each atom shares a valence electron with four neighbouring atoms, two in the plane above and two in the plane below, in order to obtain the requisite total of eight. However, since there is not a one-to-one correspondence between the top layer of silicon atoms and the adjacent layer of germanium atoms, some of the silicon atoms are unable to acquire a full complement of electrons. This is illustrated in Figure 8.4(a), where it can be seen that the silicon atoms labelled X and Y form bonds with only three other atoms. This is of great importance since these atoms are likely to trap any stray conduction electrons in order to complete their unfilled shell. This renders the material completely useless from either an electronic or an optical point of view.

Fortunately, the atoms in a crystal do not tend to behave in this way Let us look at the tennis ball model again. If we exert some force on one of the larger balls, we can distort it slightly so that its size along two of the directions is reduced. This suggests that if we squash the balls together it may still be possible to fit one hundred of the larger balls in each row. Each ball can now occupy the desired position in a plane which has the same spacing as each of the planes below. If we grow a thin layer of germanium atoms on to the surface of a silicon crystal an analogous effect occurs. The germanium atoms pack themselves together more tightly than normal in order to fit into the appropriate places above the topmost plane of silicon atoms. This means that in two directions they have the dimensions of a silicon atom, and as a result they bulge outwards in the direction perpendicular to the plane. Since the germanium atoms are distorted, we say that this atomic layer is strained. Figure 8.4(b) gives some idea of this behaviour. Again the distortion is exaggerated for clarity.

Although thin layers of atoms naturally arrange themselves in these strained configurations, there are severe restrictions on the thickness of these layers. We can see why by considering the following argument. The first layer of germanium atoms to be grown on the silicon surface is compelled to adopt the same spacing as the silicon atoms, even though this means that they are more closely packed than they would normally choose to be. If we now form a second layer of germanium atoms, although these are placed directly on top of other germanium atoms, the spacing between the atoms is still dictated by that of the silicon atoms in the plane below. As further layers of germanium atoms are grown, each one is forced to adopt the spacing imposed by the uppermost silicon plane, which becomes increasingly distant with each successive layer. There is a very delicate balance here. The further the germanium atoms are removed from the silicon plane, the more likely they are to rebel against the oppression of this silicon regime and demand the space to which they are entitled. When this happens the crystal structure is suddenly destroyed, and fault lines, called dislocations, spread through the germanium layers.

The relaxation of the strained layers occurs typically when a certain critical thickness is reached. In cases where the discrepancy in the atomic spacing is only one percent or so, about fifty atomic layers can be grown before exceeding this limit. However, the comparatively large difference of four percent means that only six atomic layers of germanium can be grown on a silicon surface. How can layers of this thickness be grown accurately? The gas deposition method described in Chapter 6 is quite adequate for producing layers about a hundred atoms thick, where we are not really concerned if we end up with a layer which is actually ninety-eight or one hundred and two atoms thick. However, it is not suitable for producing thicknesses of only six atoms, particularly when we require many identical layers to produce a superlattice. Instead a more precise technique is required. The preferred method is usually MBE, an acronym for molecular beam epitaxy.

An MBE machine is an enormously expensive piece of apparatus costing in excess of a million pounds. For twenty years it has represented the state of the art in the production of layered atomic structures with individual layer widths ranging from over a hundred atoms to only a couple of atoms thick. Its basic principle of operation is as follows. Suppose that we take a sample of some material, such as silicon, and place it in a closed container. We will refer to this as an oven since we are going to heat the silicon so that some of the atoms break free from the crystal and form a gas. Gas molecules move about randomly, colliding with each other or with the walls of the oven. If we now make a very small hole in the oven, then some of these atoms will be able to escape. It will appear as though the oven is emitting a narrow beam of silicon atoms. This is where the 'molecular beam' part of MBE comes from, although in this case the 'molecules' are single atoms of silicon. A piece of silicon crystal, called the substrate, is placed in the line of the beam, as shown in Figure 8.5. This forms the foundation for the growth of subsequent layers. When it is irradiated with a beam of silicon atoms, the atoms stick to the surface and new layers are grown on top of the existing crystal. The major attraction of MBE is that the growth rate is very slow—typically only one plane of atoms is formed per second. We can easily and very quickly halt the growth of the crystal at any stage by simply sliding a shutter across the hole in the oven. Since this operation can be performed in a fraction of a second, we have the ability to control the thickness of the layer with atomic precision. Once the silicon beam is shut off we can open the shutter on another oven containing germanium, say, and proceed to grow layers of germanium atoms on top of the silicon surface. As we discussed previously, the germanium atoms arrange themselves in appropriate places so

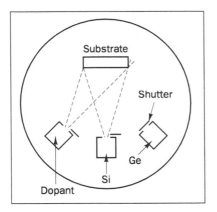

Figure 8.5 A simplified schematic diagram of an MBE machine.

as to extend the structure already present in the substrate. After a few seconds the germanium shutter is closed again, and the silicon one reopened. By using a computer to control the opening and closing of the shutters, a large number of identical alternate layers can be grown with extremely high precision.

One potential problem concerns the smoothness of these layers. Since the atoms hit the substrate at random positions, it is to be expected that the surface of these layers will be quite rough on an atomic scale. For example, if we attempt to grow six atomic layers of germanium we might expect to have some regions where a thickness of, say, eight atoms has accumulated, and other regions with a thickness of only four atoms. Growing further layers of silicon and germanium on to such a surface would produce rather wavy layers of varying thickness. We can solve this problem by keeping the substrate at a high temperature, typically between four hundred and fifty Celsius and six hundred and fifty Celsius. This ensures that the atoms arriving at the surface have sufficient thermal energy to enable them to move around. As they do so they tend to settle in places to form planes which are atomically smooth over large areas. To take full advantage of this mechanism it is often best to leave a short interval (of a few seconds) between ceasing the growth of one type of material and starting the growth of another. This allows the atoms on the surface enough time to rearrange themselves to produce as smooth a layer as possible.

How can we tell how smooth the surface is? Fortunately, it is possible to monitor the roughness of the surface as the layers are being grown. This is achieved by using a beam of high-energy electrons directed at a very shallow angle on to the surface where growth is taking place. The results are easy to interpret if we consider an analogous process carried out on a more familiar scale. Suppose we fire a number of tennis balls so that they strike a surface at a glancing angle. On a smooth surface the balls bounce off in a predictable manner, whereas on a rough surface they scatter in different directions. The electrons behave in a similar way, except that they are capable of indicating roughness on an atomic scale. Since these measurements are performed within the growth chamber, changes can be made during growth to improve the quality of the layers.

A further vital aspect of producing semiconductors for use in devices is the need to dope the material. In Chapter 5 we found that silicon chips usually start life being uniformly doped with one particular type of doping. Regions with opposite types of doping are then created by further processing. For example, p-type regions can be formed within an n-type semiconductor by either allowing acceptor atoms to diffuse in to the surface of the material at high temperature or by literally firing the acceptors at high speed in to the crystal. The volume of the crystal which is affected by this doping can be controlled within a reasonable degree of accuracy by using masks to define the areas which are to be doped, and using the many years of accumulated knowledge to determine how far the dopant atoms penetrate into the crystal. In contrast, MBE allows extremely precise doping, and has the ability to include both types of dopant atom during growth. All that is required is two further ovens within the MBE chamber, one containing donor atoms and the other acceptor atoms. Let us see how we could introduce doping as we grow a crystal of gallium arsenide. Since the gallium and arsenic sources are extremely pure we produce a crystal with virtually no impurities. We previously introduced the term 'intrinsic semiconductor' to describe such a material. If the shutter on the oven containing the donor atoms is opened very briefly, then these impurities will dope the surface layers of the growing crystal. Once the shutter is closed, further layers of intrinsic gallium arsenide are grown on top, leaving the donor atoms contained in a layer of the crystal just a few atoms thick. Such precision is of enormous advantage in the construction of MODFETs and similar structures where we wish to isolate the moving conduction electrons from the impurity ions. It also enables us to produce an entirely new kind of superlattice structure.

So far we have considered superlattices formed from alternate layers of two different materials. These are referred to as compositional superlattices. A further type of superlattice, with quite different properties, can be formed by using a single material, such as gallium arsenide, and doping thin layers alternately with donors and acceptors. It is usual in such systems to separate the doped layers with regions which are undoped, and the alternate layering of n-type, intrinsic, p-type, intrinsic gives rise to the name 'nipi' superlattice.

Before we look at the properties of a nipi superlattice, let us first concentrate on a small section of such a system. Suppose we focus on a thin layer of n-type material somewhere in the middle of this system. On either side there is a layer of intrinsic material, followed by a thin layer containing p-type doping. Although the entire structure is only a few hundred atoms across, far smaller than the depletion region of a p-n junction, we can get some idea of what happens by remembering how the carriers are redistributed at such an interface. Let us first concentrate solely on the conduction electrons. We can imagine that these are

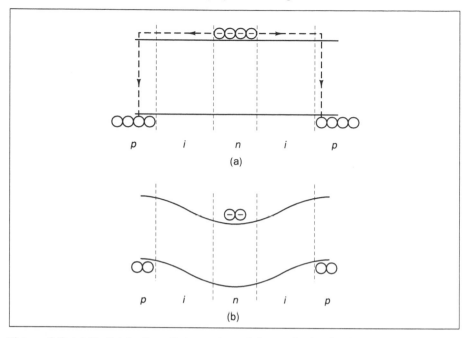

Figure 8.6 (a) Redistribution of charge in a nipi superlattice leads to (b) a significant change in the band edges, reminiscent of that in a p-n junction.

initially confined to the narrow n-type layer, as in Figure 8.6(a). However, the principle of diffusion ensures that they do not stay there for long. The electrons move outwards, and as they do so, some will reach the p-type layers and recombine with holes. This makes the p-type region negatively charged, and at the same time the donor ions which are left behind in the n-type layers render these regions positively charged. This too is reminiscent of the p-n junction. We can see from Figure 8.6(b) that as further electrons move across to the p-type layer, it becomes increasingly difficult for the electrons to leave the n-type region. Another way to view this is to suppose that we are digging a trench in the ground. Initially it is quite easy and we place the soil on the ground at either side, but as we dig down the work gets progressively harder. It takes increasingly more effort to remove the soil because as the trench gets deeper, the piles of earth on either side grow higher. Eventually we would probably give up when the difference in height becomes too great. Using this analogy we see that as the electrons are transferred to the p-type regions, those still contained in the n-type layers need increasingly more energy to escape. Finally, we reach a point where the electrons can no longer manage to escape. They are trapped in this region just as effectively as the electrons in a gallium arsenide layer sandwiched between two algas layers. If the region is small enough—in other words, if it is comparable in size with the wavelength of an electron—then quantum effects will be important. This means that we can create a quantum well simply by precise doping of a single type of material.

A similar argument applies if we consider the holes in the valence band. Some of the holes diffuse outwards from the p-type layers and recombine with electrons in the n-type layer. The difference in charge also produces a quantum well to trap the holes. This again appears to be similar to the gallium arsenide quantum well. However, there is a distinct difference. In this case the electrons are confined in one region (the layer doped with donors), while the holes are confined in different regions (those doped with acceptors).

Let us examine the energy gap of this system. We have seen that this is an important property since it determines the wavelength of the light which can be absorbed or emitted by electrons moving from one band to another. The energy gap of a compositional superlattice can be tuned by varying the thickness of the well and barrier and the composition of the layers. In a similar way, the electron and hole energies in a nipi superlattice are determined by the concentrations of the dopant atoms and the widths of the doped and intrinsic regions. Of course, all of these parameters are fixed once the structures have been fabricated. In a compositional superlattice it is then no longer possible to adjust the band gap of the system. The remarkable feature unique to nipi superlattices is that even after they are constructed, it is still possible to control the energies of the electrons and holes. To see how this is achieved we need first to examine how electrons and holes recombine in these systems.

We know that conduction electrons tend to lead a rather precarious existence. If such an electron comes into close proximity to a hole in the valence band, the electron jumps down into this lower energy state. In most materials, the lifetime of a conduction electron, which is the average amount of time that an electron exists in the conduction band before recombining with a hole, is typically measured in millionths of a second. In a nipi superlattice the situation is very different. The electrons and holes are separated from each other, being confined in different regions of the crystal. This suggests that the carriers should have a virtually infinite lifetime, recombination being possible only if a conduction electron gains enough thermal energy to move into one of the higher energy conduction states in the p-type region. However, there is another way in which electrons and holes can recombine. This is achieved if a conduction electron tunnels out of the well in the n-type layer into the neighbouring p-type layer, from where it can then recombine with a hole. There is a conceptual difficulty with this picture in that when the electron tunnels into the p-type region it ends up at a forbidden energy level in the crystal. In order to bypass this problem we can once again invoke the uncertainty principle. The process is illustrated schematically in Figure 8.7. We can think of the electron borrowing energy so that it moves into one of the allowed conduction states in the p-type region. The electron can only remain in this state for a very short period, after which time

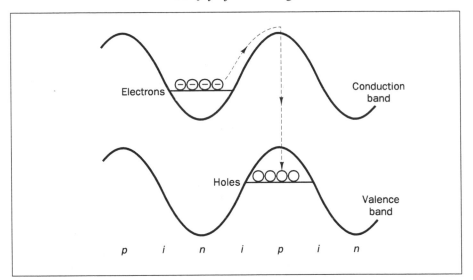

Figure 8.7 In a nipi superlattice the conduction electrons and holes are confined in different regions. A conduction electron must first borrow some energy in order to recombine with one of the holes and reach a lower final state.

the energy loan must be repaid and the electron returns to its original position. However, if the electron encounters a hole within this time period, then recombination takes place. Since the hole has a lower energy than that initially possessed by the electron, the energy loan is repaid in full, and the remaining surplus energy is emitted as a photon.

Using this picture we can also predict that the rate at which recombination occurs will be dependent on the depth of the well. The argument goes as follows. According to the terms of the uncertainty principle, the time period over which the energy loan must be repaid becomes shorter as the size of the loan increases. If the quantum well is deep, then the energy required to move into the p-type region is greater than that for a shallow well. The electron therefore spends less time in the vicinity of the holes, and so the probability of recombination occurring is reduced. We conclude that electrons in deep wells have greater longevity than those in shallow wells.

Let us now look at how we can interactively tune a nipi system. Consider a system which is in equilibrium. We will assume that the quantum wells are relatively deep, as in Figure 8.8(a), so that the resultant band gap is significantly smaller than that of the host semiconductor. We now illuminate the crystal with a beam of light. Each photon is capable of creating an electron–hole pair. The electrons naturally collect in the lowest-energy regions of the conduction band, that is in the n-type layers, and for similar reasons the holes collect in the p-type layers. At this point we should remind ourselves that the wells and barriers

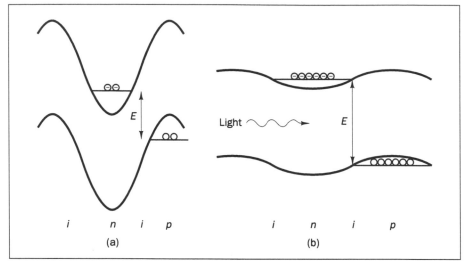

Figure 8.8 Demonstration of the interactive tuning capability of a nipi superlattice. Illumination by a beam of light changes the system from (a) to (b) producing a larger band gap, E, a greater concentration of carriers and shallower wells and barriers.

appear in this structure only because of the redistribution of charge. Equilibrium was achieved because electrons migrated to the p-type layer and recombined with holes, leaving behind a net positive charge in the n-type layers and producing an overall negative charge in the p-type layers. The effect of shining light on to the material is therefore to go some way to restoring the balance. The wells become shallower, as can be seen from Figure 8.8(b). This in turn makes it easier for the electrons and holes to recombine, and so some of the electrons are transferred back into the p-type layers. A new equilibrium condition is achieved when the rate at which the electrons recombine with holes is equal to the rate of creation of new electron–hole pairs by the photons in the light beam. The interesting point is that by illuminating the system we have changed several of its properties. The energy gap of the system, the number of conduction electrons and the number of holes have all been increased, as we can see from Figure 8.8(b). In addition, the decrease in the depth of the well leads to a corresponding decrease in the lifetimes of the carriers. We should also note that this is a reversible process. When the light beam is switched off, recombination of the electrons and holes will become the dominant process for a brief moment until the original equilibrium condition is restored (as in Figure 8.8(a)).

A simple illustration will serve to show how we could make use of this tunability. Suppose that the crystal is illuminated with light of a single wavelength—that is, the photons all have the same amount of energy, which we assume is somewhat greater than the energy gap. If this light is of low intensity, then the rate at which electrons are excited into the conduction band is slow and the energy gap is not greatly altered. However, we should note that the light which is emitted as these electrons and holes recombine is of a wavelength corresponding to this new band gap. Now let us consider what happens if we increase the intensity of the light. Electrons and holes are created at a greater rate, and, as we have just seen, the band gap increases. As the electrons and holes recombine the change in energy is now greater than before, and so the emitted light will be of shorter wavelength. This is quite an interesting phenomenon. By keeping the wavelength of the incident light beam constant, but changing the intensity of the beam, the nipi superlattice modulates the wavelength of the emitted light.

We will return to the subject of using materials to influence the behaviour of optical signals in Chapter 11. In this chapter we have seen how novel materials can be produced by growing crystals in which alternate layers of two different types of material, or two different types of doping, are carefully arranged. These new crystals can have properties quite different from those of any naturally occurring materials. As we shall see in the next chapter, they also have the potential to form the basis for extremely fast-switching devices.

9

Negative Resistance and the Quantum Transistor

IN the previous two chapters we have seen how quantum physics affects the behaviour of electrons confined to very small regions in one or more dimensions. In the process we have considered several optical applications of these effects. However, the initial reason why we were led to consider such small length scales was in the pursuit of ever faster transistors. Is it possible to produce transistors which exploit these quantum effects? If it is, then we can be sure that the principle of operation is likely to be very different from that of conventional transistors.

Let us return to the system of two quantum wells which we considered at the beginning of the previous chapter, and assume that all of the electrons start off in one of the wells. Initially the electrons tunnel from the full well to the empty one, and so a current flows. However, after a short time the concentration of electrons in the two wells becomes equal, and the current ceases. (We should note that the electrons continue to tunnel between the two wells, but since there are equal numbers tunnelling in each direction, the net current is zero.) This is not a very promising basis for a transistor. To build such a device we must be able to control the flow of electrons. How do we achieve this?

As a starting-point let us look at a modification of a p-n junction diode. By now we should be familiar with the fact that there is a transfer of charge at the junction with the result that the electrons in the energy levels in the n-type region end up at lower energies than the corresponding levels on the p-type side. We also know that the difference in energy depends on the amount of transferred charge, which in turn varies with the concentration of the dopants. By using very high levels of doping, it is possible to alter the energy levels to such an extent that the conduction electrons on the n-type side have less energy than the holes on the p-type side. Such a situation is illustrated in Figure 9.1(a). The high concentration of doping means that the depletion region around the junction is very narrow, which in turn means that there is only a very small spatial separation between the electrons in the conduction band and the holes in the valence band. However, it is not favourable for the electrons and holes to recombine since the electrons have a lower energy if they stay where they are

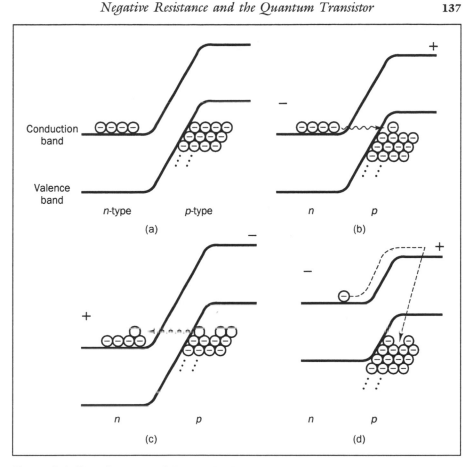

Figure 9.1 The alignment of the conduction and valence bands in a tunnel diode (a) with no applied voltage, (b) with a small forward bias, (c) with a small reverse bias, and (d) with a large forward bias.

in the conduction band than they would have if they were to enter one of the vacant valence band states. Consequently, an equilibrium condition exists in which a substantial number of electrons are present in the conduction band.

What happens if we apply a voltage to this system? Suppose that we apply a small forward bias, making the n-type side more negative with respect to the p-type side. This reduces the energy differences at the junction, as shown in Figure 9.1(b), and so the conduction electrons are raised in energy. These electrons now have a slightly higher energy than some of the hole states in the valence band on the p-type side. This makes it very attractive for electrons to tunnel from the n-type side into the valence band on the other side, and so a current flows across the junction. The result appears to be the same as the result we obtained with a normal p-n junction diode under forward bias, although the mechanism by which the electrons recombine with the holes is quite different. In view of this, we will refer to this structure as a tunnel diode.

What if we now reverse the electrical connections to the diode, so that the p-type side is made more negative? This increases the energy difference between the n- and p-type layers, and so in the ordinary diode makes it harder for the electrons and holes to recombine. However, if the polarity of the connections to the tunnel diode are reversed, the current continues to flow, but in the opposite direction. It is easy to see why with the help of Figure 9.1(c). There are now valence electrons on the p-type side which have more energy than some of the vacant states in the conduction band on the n-type side. The electrons therefore tunnel across to fill these lower energy states, and so the current flowing across the junction is in the opposite direction to that obtained with forward bias. This idea takes a little bit of getting used to. It seems highly irregular that electrons should be able to flow so freely from the valence band into the conduction band, but this is merely a consequence of the rather strange alignment of the bands. The result is slightly worrying for another reason. We expect a diode to permit the flow of current in only one direction. In fact, we saw previously that the asymmetric behaviour of the p-n junction is vital to the operation of the transistor. However, in this case we appear to have a diode which permits current to flow freely in either direction!

Fortunately this is not quite the full story. So far we have examined only what happens when very small voltages are applied. If a larger forward bias is used then we discover that the tunnel diode exhibits a quite startling feature. When subject to a small forward bias the electrons tunnel across the junction, but if a slightly larger bias is applied, the electrons on the n-type side are raised higher in energy than the vacant states in the p-type side (Figure 9.1(d)). The electrons can no longer tunnel directly into the adjacent layer, since no corresponding energy states exist on the other side of the junction. Instead they have to use the mechanism that we discussed in the previous chapter when considering the nipi superlattice. An electron can use the uncertainty principle to borrow the required energy so that it can move up to the conduction band on the p-type side. From there it can recombine with a hole, assuming that it encounters one within the time limit of the energy loan. As we showed previously, the rate at which such events occur is very small when the difference between the energies of a conduction electron on the two sides of the junction is large. In particular, the rate at which electrons cross the junction by this process is many times smaller than that when the electrons can tunnel directly from one side to the other. This means that the small increase in voltage required to take us from the situation in Figure 9.1(b) to that in Figure 9.1(d) leads to a dramatic decrease in current. To appreciate just how strange this behaviour is, let us examine the relationship between the voltage and current in an ordinary material.

We have seen that the effect of applying a voltage to a crystal can be likened to that of tilting the energy bands, causing the electrons to move down towards the lower, positive end. Electric current, on the other hand, is a measure of the velocity at which the electrons travel towards the positive end of the crystal. In the simplest picture it would seem that an increase in voltage will produce a greater degree of tilt and so give rise to a corresponding increase in current. This relation was first stated by the German physicist Georg Ohm in 1826. He noticed that in many cases the current varies linearly with the voltage, the constant of proportionality between the two quantities being referred to as the resistance of the material. This is now known as Ohm's Law. As we discussed in Chapter 6, the electrons do not all rush towards the positive terminal as this picture suggests. Their motion is far more random, with just a slight tendency to move in the direction of the applied field. Nevertheless, a detailed analysis of such motion shows that, in general, this model does predict a linear relationship between current and voltage, in agreement with Ohm's Law. There are, of course, many cases in which Ohm's Law is not satisfied. For example, when large voltages are applied the current tends to be rather smaller than that predicted by Ohm's Law. We notice an analogous effect when driving a car. As we push down on the accelerator pedal, the car travels faster, but as we approach the car's maximum speed, it becomes increasingly difficult to make the car go faster. More drastic deviations from Ohm's Law are encountered in many electronic devices, such as the p-n junction and the transistor. However, although there is no longer a linear relationship, the observation that an increase in voltage produces some increase in the current always remains true.

The tunnel diode displays quite different characteristics. As the forward bias is increased, the current first increases and then decreases rapidly. To return to the analogy with the car, it is as though we depress the accelerator further only to find that the car slows down! Since electrical resistance is a measure of the increase in currrent with voltage, a decrease in current can be interpreted as a negative resistance. This phenomenon is illustrated in Figure 9.2.

Without going into great detail we will briefly mention one application of this negative resistance in an analogue device. It will suffice to say that when the tunnel diode is operating in the region of negative resistance it can draw power into an electrical circuit in which it is placed. In addition, as the depletion region of this device is very narrow, the transit time for an electron to cross this region is extremely short. Consequently, the device can be used to create or amplify rapidly varying electrical signals in the microwave range. However, our main interest in this book is with digital electronics, and we will see shortly how a negative resistance can be useful in this context. First, though, we need to consider how the tunnelling ability of electrons can be exploited in a transistor.

We have seen that when quantum properties, such as tunnelling, become important in the operation of a p-n junction it behaves in a quite different

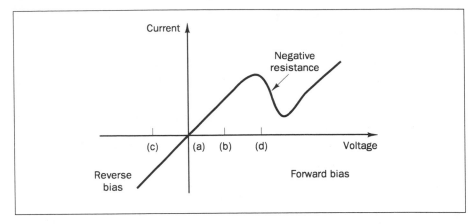

Figure 9.2 A typical relationship between voltage and current for a tunnel diode. The letters (a)–(d) correspond to the cases shown in Figure 9.1(a)–(d).

manner from what we would expect in a classical case. Clearly, if we wish to construct a transistor which employs tunnelling, it is not appropriate simply to scale down a conventional device. Instead we need to take a rather different approach.

Many of the designs for quantum transistors which are currently under development make use of the concept of resonance. We have all experienced resonant effects. For example, when travelling in a car it is often found that the engine noise becomes particularly noticeable at a certain speed. This is because the wavelength of the sound becomes comparable with the internal dimensions of the car so that a standing wave forms (just like the wave on the stretched string that we discussed in Chapter 7). A similar effect accounts for the ability of some opera singers to be able to shatter a glass when they hit a particularly high note. In the present instance we need to consider a slightly different type of resonance. To describe this effect let us consider a building consisting of two towers, one of which is built on the east bank of a river and the other on the west bank. Let us suppose that we are in the east tower and want to get to the west tower. We start off on the ground floor, but there appears to be no way of reaching the other half of the building from this level. Even if we go out of the building we still have to get across the river. Never mind: we can see if the situation improves as we go to higher floors. On reaching the first floor we find that the situation is quite different. There is a passageway joining the two halves of the building, and so this enables us to cross quite easily into the west tower. Our curiosity is now aroused and so we decide to explore the higher floors. We find that the second and third floors do not possess a connecting passageway, but the fourth floor does. We can conclude that the building exhibits certain resonant levels at which it behaves quite differently from all the other levels. In particular, a person on the first or fourth floors can easily move from one side of the building to the other, whereas at any other level the person is confined to just one half of the building.

The reason for this diversion into architectural curiosities is that by fabricating a semiconductor device which behaves in a similar way we can produce some very interesting results. We can make such a structure by starting with a crystal of gallium arsenide. A quantum well can be formed by growing two thin layers of algas, separated by a similar layer of gallium arsenide, and a further layer of gallium arsenide is then grown on top.

Suppose we apply a small voltage to the system, making the connections to the gallium arsenide layers on either side of the quantum well. We might expect the electrons on the negative side to move towards the positive terminal, but there is a barrier in the way. To be precise, there are two barriers, separated by a quantum well. Surely these two thin layers of algas will pose no obstacle to an electron. It can simply tunnel through them, can't it? Actually, it isn't quite that simple.

Let us consider an electron in the lowest conduction state on the negative side. Can it tunnel through the first barrier into the quantum well? The answer is no, because there is no corresponding energy state in the well. As we have seen before, the lowest pernmissible energy level in a quantum well is somewhat above the bottom of the well. This means that if the electron is going to be successful in reaching the other side then it must tunnel through the first algas layer, the quantum well and the second algas layer all in one go. Since the probability of an electron tunnelling through a barrier decreases dramatically as the thickness of the barrier increases, the likelihood of such an event occurring is small. However, we know that at room temperature not all of the conduction electrons are in the lowest energy level in the conduction band. A small proportion of them will occupy some of the higher states. In most of these levels the probability of an electron tunnelling through the barriers is as small as it is for an electron in the lowest conduction state, but there is one exception. If an electron has precisely the same energy as that of the discrete state in the quantum well, then the electron can easily tunnel through the algas barrier into the well. It no sooner enters the well than it tunnels through the second barrier and pops out on the other side. The electron therefore passes from one side to the other as though the barriers were not there! We refer to this as resonant tunnelling. You may find Figure 9.3 helpful to visualize this process. It is apparent that this state of affairs bears a certain similarity to the building that we described earlier.

We can see from the above discussion that the ability of an electron to pass through this structure depends critically on the energy of the electron. One way to exploit this is to use the applied voltage to control the number of electrons which can take part in resonant tunnelling. In order to make the explanation simpler let us assume that all of the electrons occupy the lowest energy state in

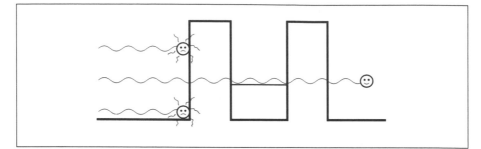

Figure 9.3 A double barrier structure containing a quantum well. Electrons with an energy corresponding to the quantum state in the well can tunnel through the structure with ease.

the conduction band. The only other piece of information that we need to know is how the energy levels change when we apply a voltage to the system. From previous discussions we know that the energy of the electrons on the negative side is shifted upwards whilst that on the positive side is shifted downwards. What about the energy level in the quantum well? Since no electrical connection is made directly to the quantum well we can assume that this remains unchanged.

Now, suppose that a small voltage is applied, as in Figure 9.4(a), so that the energy of the conduction electrons on the negative side is still lower than the energy state in the quantum well. Resonant tunnelling does not occur and so no current flows through the structure. If the voltage is increased so that the situation shown in Figure 9.4(b) is achieved, then the electrons are able to pass through the barrier structure with ease and a current flows.

However, if we increase the voltage beyond the resonance point, then the conduction electrons on the negative side are moved to a higher energy than the state in the well. Once again we find that the electrons cannot tunnel into the quantum well because there is no allowed state with the same energy, and so the current flowing through the device drops abruptly. This is just like the phenomenon observed with the tunnel diode. An increase in the voltage has produced a decrease in current, and so we have a region in which the device behaves as though it has a negative resistance. This is not quite the end of the story. Since it is quite possible to have more than one discrete energy state in the quantum well, this means that as the voltage is increased further we may reach another resonance energy, followed by another region of negative resistance, and so on. An example of the relation between the voltage and current for a quantum well containing three discrete states is shown in Figure 9.5.

Although this is potentially a very useful characteristic we should note that the device has only two electrical connections. Just as in the p-n junction and the tunnel diode it is the size of the voltage between these two terminals which

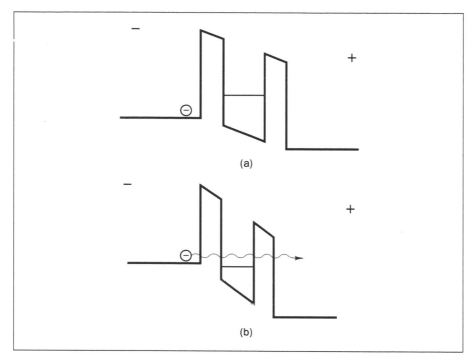

(a)

(b)

Figure 9.4 Schematic diagram of the conduction band edge in a double barrier system. In (a) the applied voltage is small and the electron cannot tunnel through to the positive side. In (b) the applied voltage is large enough so that the electron energy on the negative side is equal to the energy of the state in the quantum well.

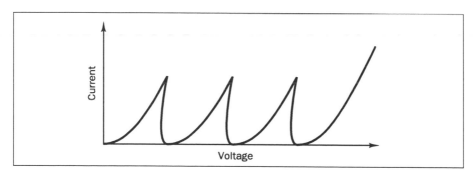

Figure 9.5 The response of a resonant tunnelling transistor which has three discrete states in the quantum well.

affects the current between the same two terminals. Such a device is not suitable for constructing logic circuits for use in a digital computer. What is required is a transistor-like device which makes use of a third terminal to control the flow of current between the other two.

Can we make a transistor which exploits the properties of the double barrier structure? Fortunately, the answer is yes. The simplest solution is to attach an electrical connection directly to the quantum well, but there are many technical difficulties with this approach. A more practical design is to incorporate the structure between the emitter and the base of a bipolar transistor. We now have a transistor which behaves under certain conditions as though it has a negative resistance. Let us have a look at some of the ways we can exploit this negative resistance, starting with a resonant tunnelling transistor in which the quantum well has just one bound state. We use a circuit configuration similar to that for a conventional transistor in a logic circuit—the supply voltage is connected to the emitter and to the output, the input is connected to the base and the collector is earthed. From the above discussion it is quite straightforward to see how the transistor operates. If the input voltage is such that the electrons in the emitter have the same energy as the state in the quantum well, then it is in a resonant condition. The current flows to earth, and the output is low. On the other hand, no current flows through the transistor if it is in a non-resonant condition, and so the output is high. Now, let us suppose that the input signal alternates in a regular fashion between a low voltage and a high voltage, the highest value being somewhat beyond the resonance level of the system. As the input voltage goes from low to high, the system starts off in a non-resonant state. Accordingly, the output changes from high to low and back to high again. Similarly, as the input falls from high to low, the output again goes through a full cycle. So every time the input voltage describes one complete cycle, from low to high and back to low, the output passes through two cycles. For obvious reasons, such a device is called a frequency doubler.

The same effect can be used to produce a logic gate from a single resonant tunnelling transistor. To do this we need to have two separate inputs to the one device. The input voltage levels are adjusted so that the resonant condition is achieved if just one of the inputs is high. In this case a current flows through the device and the output is low. However, if both the inputs are low, or if they are both high, then no current flows through the transistor and the output is high. In the latter case this occurs because the transistor is operating in the negative resistance region. Consequently, the device produces a high output if the inputs are the same and a low output if the inputs differ. Such a system is called an XNOR or eXclusive-NOR gate. (This rather convoluted name can be explained as follows. We have already met the eXclusive-OR gate in Chapter 4, where we found that it produces a high output only if the inputs are different, i.e. one or the other, but not both, of the inputs is high. An eXclusive-NOR gate simply exhibits the opposite characteristics.) What is most impressive is that we have constructed this logic gate using only one resonant tunnelling transistor. To perform the same function requires at least eight conventional transistors.

We can extend these ideas further by using quantum wells with more than one bound state. Let us consider a case with two such states and four separate inputs. By carefully designing the quantum well it is possible to arrange for resonance to be achieved with the first level if just one of the inputs is high, or with the second level if any three of the inputs are high. (Conversely, if two, four or none of the inputs are high we assume that the device is off-resonance.) This system is called a parity detector since the output depends upon whether there is an even or odd number of high input signals. These rather specialized circuits are used to check the consistency of data as it is transferred either within a computer, or from one computer to another. Twenty-four conventional transistors are required to produce a parity detector with four inputs, but only one resonant tunnelling transistor is required. And of course, if further levels exist in the well, then even more inputs can be used with a single device.

As a final example, let us look at how a resonant tunnelling transistor could, in principle, be used as a memory element. We have seen in Chapter 4 that a conventional transistor responds when a voltage is applied, but returns to its original state once the voltage is removed. We can picture this by imagining that we start off with a ball at the bottom of a slope. The height of the ball is representative of the current flowing through the transistor, so if we apply a voltage we are able to push the ball up the slope, the ball rising higher as we increase the voltage. However, when the voltage is removed, the ball rolls down again and we are back at the starting-point. As a result of this behaviour we need quite a complex arrangement of transistors to store a single item of binary information. Typically this is achieved with a flip-flop circuit (see Figure 4.9). In a resonant tunnelling transistor the response is very different. As we apply a voltage, the ball rises steeply until we reach the resonant level. Beyond this point it begins to move down the other side of the hill. If we now remove the voltage the ball does not return to its starting-point. Instead it drops into this new valley. In this way the system 'remembers' what has happened to it, and so we can use a single resonant tunnelling transistor to produce a memory element. If the quantum well has a single bound state this gives rise to a system with two stable states. If other energy levels are available then each one provides a further stable state for the system. For instance, if the well has three discrete energy levels, as in Figure 9.5, we have a memory element which works in base four. The ability to replace an entire flip-flop circuit with a single device, coupled with the use of several stable states, suggests that very high densities of memory storage could be achieved with these devices.

It is clear then that these resonant tunnelling transistors are far more than just switching devices. To see how they fit into the scheme, let us briefly review how far we have come. We have followed a path which has led us to consider transistors of ever decreasing size and then used various tricks to further increase the switching speed of these devices. However, in all cases we have been dealing with a device which has the same essential characteristic of operation, namely that it has two output states which we have variously labelled ON and OFF or TRUE and FALSE. This one device has then been used to perform a multitude of tasks including binary arithmetic, logic and memory functions. However, the implementation of quantum transistors suggests a departure from this approach. As we have seen, it is not just the decrease in size or increase in speed of the individual devices which is important, but the fact that these new devices can be engineered to perform specific functions, tasks which require many conventional transistors. Accordingly there is a change in emphasis when we assess the importance of these devices. We should no longer be concerned with the size and switching speeds of the individual devices. Instead we should be thinking in terms of the area of the chip and the amount of time required to perform a given function. We have touched here on a few examples. In the future we are likely to find an increasing variety of highly specialized devices on a single chip.

*

Having made this point, let us now turn our attention again to the physical size of the individual devices. The resonant tunnelling transistor that we have described is a vertical structure, that is, the electrons move in a direction perpendicular to the surface of the substrate as they travel from the emitter to the collector. So far we have considered only the dimensions of the device in this direction. However, if we want to place as many such devices as possible on a given surface area of the substrate then we need to minimize the area that each device occupies. In other words we have to reduce the lateral dimensions of the structure. This can be achieved quite easily using techniques familiar from the fabrication of integrated circuits. A layer of resist is formed on the surface of the structure and this is patterned using an electron beam. The layered structure is then selectively etched away in the unexposed regions leaving just one small column. In fact, with sufficient care it is possible to produce an array of several columns from the original structure. The finished article is shown schematically in Figure 9.6. Each of these short stubby islands projecting from the surface of the wafer contains a quantum well, and so is capable of acting as a resonant tunnelling transistor in its own right.

In our discussion of the resonant tunnelling transistor we have assumed that the wavelike nature of an electron in the quantum well is important only in the direction perpendicular to the well. The electrons are free to move in the plane of the quantum well. However, as we reduce the lateral dimensions of this structure we will at some stage reach a point where the wavelike nature of the

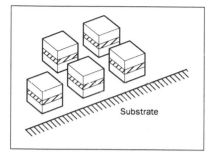

Figure 9.6 An array of resonant tunnelling transistors can be created by etching deep channels through a layered structure containing a quantum well. The free electrons are in the shaded bands. If the lateral dimensions of these islands are made small enough the quantum wells become quantum dots.

electrons in these directions also becomes important. In this case the electrons are confined in all three dimensions. This is a quantum dot, the Holy Grail of quantum electronics.

What length scales are required to produce a quantum dot? We have seen that in a quantum well the wavelike properties of the electrons only become apparent for layer widths of about a hundredth of a micron. (A micron is a thousandth of a millimetre.) To produce a quantum dot we need to achieve similar length scales along the other two dimensions. Using present techniques it is relatively easy to reduce the lateral dimensions of the islands to about a tenth of a micron. With great effort further reductions can be achieved, but producing a structure one hundredth of a micron across presents an immense technical challenge. Fortunately nature is on our side.

Suppose that we etch two parallel channels through the layered materials to leave a ridge approximately one-tenth of a micron across. In the etching process the atoms on the vertical surfaces of the ridge are disturbed so that the properties of the material are changed considerably. In fact, surfaces in general have quite different properties to bulk material. Let us consider an example using a crystal of silicon. We know that each silicon atom tends to bond with four neighbouring atoms to obtain a full complement of electrons in its outer shell. However, atoms at a surface cannot always find another four atoms with which to bond. This means that surface atoms have a tendency to hold on to any nearby free electrons in order to obtain a full shell. This process alone tends to remove conduction electrons from the surface layers. In addition, the accumulation of negative charges trapped by these surface atoms creates an electric field which repels other electrons from this region. As a result we find that the only free electrons in our tiny ridge-like structure are confined to an even smaller channel in the centre. It is rather like a coaxial cable in which the central copper core is surrounded by an insulating layer. For a ridge which is a tenth of a micron across, the electrons are confined to a region which is only about a hundredth of a micron wide. This is just the right length scale required to observe the wavelike nature of the electron. So although at first sight the dimensions appear to be far too large, we can observe lateral quantum confinement in these structures.

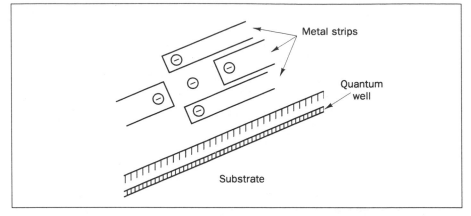

Figure 9.7 A quantum dot can be formed by enclosing an area of a layered structure between negatively charged electrodes. The conduction electrons are confined to a small region of the quantum well directly beneath the enclosed area.

In this case we have created a quantum wire, since the electrons are still free to travel along the length of the ridge. From this stage the construction of a quantum dot is relatively straightforward. If we etch another set of similarly spaced grooves at right-angles to the first set, then the surface effects in these directions are sufficient to cause the electrons to be trapped in a tiny box. Using these techniques it is possible to pattern a relatively large area to produce a regular grid of quantum dots.

An alternative method of producing quantum dots is to confine the electrons using electrodes. This concept is illustrated in Figure 9.7. We again start off with a layered structure which contains a quantum well and deposit narrow metal strips on the surface of the structure so as to define a square area. If we make all of the metal strips negatively charged then the electrons in the quantum well beneath the surface will be repelled. However, since they are confined to a thin layer which is parallel to the surface, the only direction they can move is sideways. As a result, the electrons are concentrated into a tiny cluster directly underneath the central square defined by the electrodes. If we make this area small enough we again have a quantum dot.

An interesting possibility arises in that we can control the number of electrons on this dot simply by varying the voltages on the electrodes. For instance, suppose that just a single conduction electron is contained in the dot. This bears a striking resemblance to a hydrogen atom since it has just one electron, the motion of which is constrained in all directions. However, there are some very interesting differences between an ordinary atom and the quantum dot. An atom has a spherical form, with the electron moving under the influence of a central nucleus, whereas the shape of the quantum dot is defined by the external electrodes and the barriers around the quantum well which constrain the electron. Since these can be altered, we can control the shape and size of this artificial atom. We can produce a cubic dot, a cylindrical dot, or use just three electrodes to produce a dot with a triangular cross-section. The possibilities are almost endless. And as long as there is just one electron in the dot these are all variations on the 'hydrogen' configuration. Of course, we are not limited to hydrogen. By placing two electrons on a dot we can create a 'helium' atom and study how the two electrons interact. Or we could have three or four electrons—in fact any number up to several hundred. Mark Reed of Yale University, one of the key workers in this field, has coined the phrase 'designer atoms' to describe these structures.

Designer atoms can in turn be used to produce an almost inconceivable range of new materials. For example, if several of them are placed in close proximity, the electrons on the dots interact to produce designer molecules. However, the composition of these 'molecules' is not limited by the chemical considerations we dealt with in Chapter 1. Virtually any arrangement would be possible. We can extend this idea further to produce a long string of quantum dots. The discrete electron energies interact to form bands, just as the atomic energies form bands in a crystal, or the quantum well states form minibands in a superlattice. The difference in this case is that the bands exist in only one dimension. The electrons are still confined in the other two planes, and so we obtain a truly one-dimensional crystal. Similarly, an array of quantum dots could be used to form a two-dimensional crystal.

Scientists have only just begun to explore this amazing world in the last few years. It is almost impossible to imagine what discoveries lie in store, but let us at least try to make some guesses as to how quantum dots could be used in future microelectronics. Suppose we produce an array of quantum dots using a grid of electrodes which divides the surface into squares about a tenth of a micron across. A quick calculation shows that we could place ten billion of these on an area of one square centimetre. If each one serves as a resonant tunnelling transistor, performing the function of several conventional transistors, such a structure should have a capability far in excess of the most powerful present-day supercomputer, and all of this on a single chip!

Of course, there are many problems to be solved first. For example, how do we make an electrical contact to each transistor? Even if we can solve this problem, there is the intricate network of interconnects between the devices to consider. As we saw in Chapter 5, the limitations on the speed at which the devices on a chip can be operated may ultimately be determined by the long interconnects which are necessary in any conventional chip layout. In this case the delay times associated with these interconnects are likely to far exceed the switching times of the individual devices. This is most frustrating. We have gone to great lengths to make a quantum dot transistor, only to discover that when we incorporate it into an integrated circuit we cannot make use of its tremendous capabilities. As so often seems to happen, we have reached the Holy Grail, but now we have got there we don't know what to do with it!

We can at least learn something from this. To utilize fully the potential of quantum dots calls for radical changes in the design and philosophy of integrated circuits. We have already observed earlier in this chapter that future trends are likely to be towards functional devices. We also know that as devices are placed in closer proximity to each other it becomes harder to isolate one from another. One answer then is actually to exploit the interactions between the devices. A specific example of such a system has been proposed by a group at the University of Notre Dame, Indiana. The basic unit of this construct is a cell consisting of five quantum dots, the number of electrons being controlled so that there are precisely two in each cell. Because of the repulsion between the electrons there are two likely configurations for each cell, as shown in Figure 9.8(a), which we can label as '0' and '1'. We therefore have the basis of a binary system, just as in a conventional digital computer. Neighbouring cells interact with one another because the distribution of electrons in one cell tends to induce a similar distribution in the next cell. Consequently, if we have a row of cells, as in Figure 9.8(b), and we set the value of the cell at the left-hand end to the state 1, this causes the next cell to adopt the same state, which in turn affects the cell after that, and so on. In this way a signal propagates along the row. More complex interactions arise if we have a two-dimensional array of cells, as in Figure 9.8(c). If we apply an input to the cells on one side of the array the cells within the array change state so that the system reaches a stable configuration. At this point we determine the output of the array by examining the states of the cells on the far side of the array. In this case it is meaningless to talk of individual devices: the whole array of quantum dots behaves as a single device performing a highly complex function.

These ideas all revolve around the use of quantized energy levels. By varying the energies of the electrons outside the quantum structure we can greatly alter the probability of an electron tunnelling through the structure, and so affect the flow of current through the device. An entirely different class of devices makes use of the wavelike properties of the electrons within these structures.

∗

We saw in Chapter 7 that a characteristic of wavelike phenomena is the ability to interfere. In particular, there are two extremes represented by constructive and destructive interference, which we could associate with the ON and OFF states of a transistor. The problem is, how do we apply this to an electronic device?

Let us first return to the familiar example of water waves. We can picture a

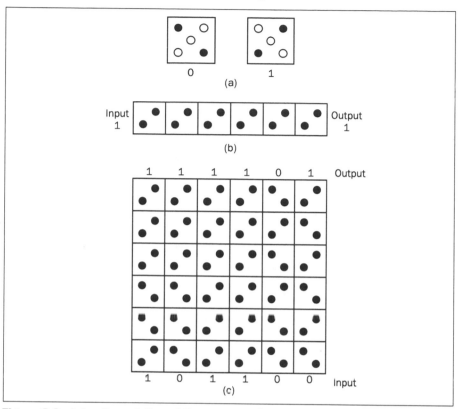

Figure 9.8 a) A cell consisting of five quantum dots containing two electrons has two stable states. Filled circles represent a quantum dot containing an electron and open circles represent a dot without an electron. b) Interactions between cells cause a signal to propagate from one cell to the next. For clarity only those dots containing an electron are shown. c) An array of cells could be used to perform a complex function by applying an input signal to one side of the array and reading the output from the opposite side.

series of regularly spaced water waves travelling along a narrow channel. Suppose that the channel divides into two, and that these two parts rejoin after a short distance. What happens to the water waves when they recombine? Let us answer this by following a specific wave as it moves through the system. As the channel splits into two, the wave is divided and travels along both channels. If both routes are of the same length, then the wave peaks in the two halves arrive simultaneously at the point where the channels rejoin, and so we have constructive interference—the final wave is the same as it was before it divided. However, if the two alternative paths are of different length, then the wave peaks will arrive at different times and there will be some mutual cancellation of the waves. In particular, if the difference in length is equal to half the distance between two

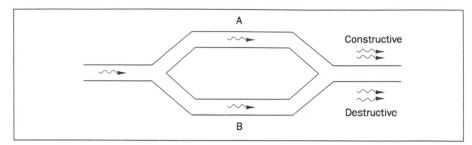

Figure 9.9 A structure for demonstrating interference of electron waves. Differences between the two paths may produce destructive, rather than constructive, interference.

adjacent wave peaks, then as the waves recombine one will be at a peak when the other is at a trough. As a result, the two will cancel out and the wave will disappear. This is a case of destructive interference.

We can produce an analogous effect with electron waves by using quantum wires arranged as in Figure 9.9 to produce a structure similar to the water channel. (Remember that in a quantum wire the electron is confined in two dimensions, and so propagates only along the axis of the wire.) Again we consider the wave dividing into two and then recombining to form a single wave. As in the example with the water waves, we obtain constructive interference if the two paths are identical, and destructive interference if they differ in length by half the wavelength of the electron. This is not a particularly useful result from the point of view of making a transistor-like device. It appears that we must physically change the length of one of the paths in order to switch between the two states. Fortunately, there is an alternative.

In 1959 Yakir Aharanov and David Bohm made a theoretical prediction that an electron wave passing near to a completely shielded magnet would undergo a change in phase. A similar effect is also predicted for an electron passing close to a totally shielded electric charge. This is conceptually very difficult to understand. If the electric charge is completely shielded so that it exerts no attractive or repulsive force on the electron, then it is hard to see how it can have any other effect on the electron. However, for our purposes we do not need to worry about whether or not the charge is completely shielded—it is sufficient to know that a change in voltage causes the phase of the electron to change. Consequently, if we return to the electron interference structure in Figure 9.9, we can see that if a voltage is applied to one arm of the device, the phase of the electron wave travelling along this path will be altered relative to that of the electron wave in the other arm. By varying the voltage we can therefore control whether the two waves interfere constructively or destructively when they recombine. In this way the magnitude of the voltage determines whether or not a current flows. We therefore have a device which has similar characteristics to

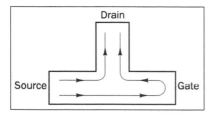

Figure 9.10 Interference of electron waves in a T-shaped structure can be considered to arise as a result of electrons following two different paths, as indicated.

a transistor. It has two distinct states and we switch between the two by applying a voltage to the structure. The major difference is in the way in which the device operates.

Another manifestation of quantum interference leads to the rather strange result that a transistor-like device can be constructed in which the gate does not lie between the source and drain. One proposal is for a T-shaped structure, as in Figure 9.10, in which the source, drain and gate form the three ends of the T. Classically, we would expect the electrons simply to flow straight from the source to the drain without being affected by the gate region. However, because of the wavelike nature of the electron, we can picture part of the electron wave taking an alternative route from the source to the drain via the gate. In this way the current of electrons arriving at the drain depends on the interference between the waves which have followed these two different paths. In order to produce a switching effect we again make use of the Aharanov–Bohm effect discussed above. If a voltage is applied to the gate it creates a change in phase of the electron wave which travels via the gate, and therefore alters the interference condition at the drain.

Although such a result seems very strange, a similar structure is commonly encountered in microwave engineering where the wavelength of the microwave radiation is typically measured in centimetres. In a similar way it may be possible to emulate many other microwave devices using electron waves, the difference being that the length scales involved are of the order of a million times smaller.

Perhaps the most intriguing possibility is an array of antidots, a structure which can be considered as being complementary to a quantum dot array. Instead of confining the electrons to energy levels within a series of quantum dots, the dots become the forbidden regions and the electrons move through the intervening channels. We could fabricate such a system by etching a regular grid of holes through a quantum well layer to produce a structure such as that in Figure 9.11. Alternatively, if we apply a positive voltage to a grid of electrodes, then the electrons are confined to a similarly shaped region in the quantum well layer beneath the surface. In each case we are left with a fine network of quantum channels through which the electrons can travel and interact with one another.

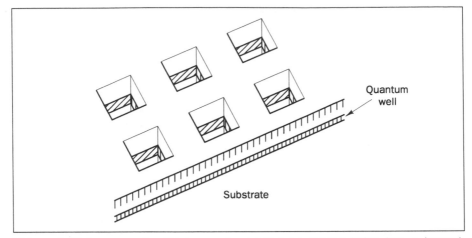

Figure 9.11 An array of antidots can be created by etching holes into a layered structure.

To summarize this chapter, we have introduced a rather bewildering range of devices, each of which is capable of performing a function similar to that of a conventional transistor. It is unlikely that all of these prototypes will make it to the market place, but some of them will almost certainly begin to appear in commercially available circuits within a few years. However, we have not yet explored all of the possible avenues. In particular, we have concentrated solely on semiconductor devices. In the next chapter we shall see how superconductors and ordinary metals can be used to produce similar types of devices, while in Chapter 11 we consider the possibility of constructing devices whose inputs and outputs are beams of light rather than electrons.

10

Superconductors and Single Electron Tunnelling

I N March 1987 the term 'superconductor' hit the headlines. Newspapers worldwide carried articles about these amazing materials which exhibit a complete absence of resistance to an electrical current. Fantastic claims—some true, others the product of wishful thinking—were made about the properties and applications of these materials. What was all the hype about? After all, superconductivity was by no means a new discovery. The sudden interest was due to the relatively high temperatures at which this zero resistance phenomenon was observed, despite the fact that these temperatures were not much above minus two hundred Celsius. To put these results in perspective we need to take a brief look at the history of superconductivity.

The Dutch physicist Heike Kammerlingh Onnes dedicated his life to the subject of cryogenics: the production and effects of extremely low temperatures. In 1908 he succeeded for the first time in liquefying the inert gas helium at a temperature of minus two hundred and sixty-nine Celsius. This was a significantly lower temperature than anyone had ever achieved before, being just four degrees above the very lowest possible temperature, absolute zero. In dealing with temperatures of this magnitude it is far more convenient to use the Kelvin scale. In this case the temperature interval is the same as on the Celsius scale, but the zero point corresponds to absolute zero, or approximately minus two hundred and seventy three Celsius. Accordingly, the temperature produced by Onnes was just four Kelvin.

The production of liquid helium presented Onnes with a unique opportunity to investigate the properties of materials at the lowest extreme of the temperature scale. In particular, one problem which he addressed was what happens to the electrical resistance of a piece of metal as it is cooled to absolute zero. Before we attempt to answer this, let us briefly remind ourselves once again of the causes of electrical resistance. One of the main factors is the movement of the ions in the solid. Accordingly, one school of thought argued that the resistance should decrease linearly with temperature, a state of zero resistance being achieved precisely at the point when the temperature reached absolute zero. However, electrons are also affected by defects in the crystal structure and the presence of impurities. Since even the purest piece of metal must contain some impurities and imperfections, another theory suggested that a small residual resistance should remain even at a temperature of absolute zero. What Onnes discovered when he immersed a piece of mercury in the liquid helium contradicted both of these beliefs. As the temperature was reduced, the resistance of the mercury sample dropped accordingly. Then just above the liquefaction point of helium, at a temperature of about four and a half Kelvin, an amazing thing happened. The resistance of the mercury sample vanished. It was not a case of the resistance falling steadily to zero. One moment the sample had a clearly measurable resistance, and the next, as the temperature dropped by a tiny fraction of a degree, the resistance simply disappeared. Further measurements showed that the change from the normal state to this new superconducting state occurred over a temperature range of about one thousandth of a degree.

This marked the discovery of superconductivity. Over the years a great many other materials—mostly metals, but also a few non-metals—have been found to exhibit superconductivity. In most cases this has required much lower temperatures, sometimes down to a fraction of a degree above absolute zero. A handful of materials which have higher transition temperatures than mercury have also been discovered. The metal niobium, for instance, superconducts at nearly ten Kelvin. By testing various alloys of niobium, the highest critical temperature, the point at which the metal changes into a superconductor, was slowly pushed up to twenty-three Kelvin.

The breakthrough came in 1986 with the announcement that a ceramic material, ironically a very poor conductor at room temperature, was capable of superconducting at temperatures of thirty Kelvin. In view of the slow progress over the previous seventy-five years, this was an amazing leap forwards. However, it proved to be just the first of an unprecedented run of discoveries. Within a few months a similar material was shown to superconduct at ninety-five Kelvin, and by 1988 the limit stood at one hundred and twenty-five Kelvin. The announcement of a room temperature superconductor was eagerly awaited.

This, then, was the cause of the excitement which swept through the academic community and the media alike. Superconductivity has always held a certain fascination because it appears to demonstrate perpetual motion. Of course, there is always a catch with any perpetual motion machine: in the case of superconductivity the failing is that an input of energy is required to produce and maintain such low temperatures. On the other hand, a material which superconducts at room temperature would surely exhibit perpetual motion. Unfortunately, no such material has yet been discovered. Nevertheless, the new breed of superconductors have surpassed one important barrier since the critical temperatures are higher than the boiling point of liquid nitrogen. This is of great significance because liquid nitrogen is much cheaper to produce and far easier to handle than the liquid helium which is used to provide the cooling for conventional superconductors.

Despite the requirements of exceedingly low temperature, the list of possible applications for superconductors appears to be almost endless. In fact, the majority of the cases that we will consider have only been constructed using the standard low-temperature superconductors. However, before we examine a few of these possibilities, we will first address a different but very important question. How do superconductors superconduct?

*

Several aspects of superconductivity indicate that it is not merely an extension of ordinary conductivity. For instance, we have seen that the presence of defects and impurities in a material would lead us to expect some resistance even at a temperature of zero Kelvin. In addition, there are other things that we cannot explain using the conventional model of electrical conductivity. For instance, why does the resistance fall abruptly to zero? There are also apparent anomalies in the types of materials which form superconductors. We have seen that the highest-temperature superconductors are in fact very poor conductors at room temperature, while, paradoxically, the best normal conductors, silver and copper, do not superconduct even when the temperature is within a thousandth of a degree of absolute zero. How can these features be explained?

It took many years to solve this problem. Although superconductivity was first demonstrated in 1911, it was not until 1957 that John Bardeen (of transistor fame), Leon Cooper and Robert Schrieffer arrived at a satisfactory explanation. The theory centres around the fact that at extremely low temperatures the conduction electrons in a material can attract one another and form pairs of electrons. This seems a very strange result. All electrons have a negative charge, and so we would expect them to repel one another. How and why do they pair up?

The 'why' is easiest to answer. It is simply that by forming a pair the electrons reduce their energy. In a similar way we have seen that two hydrogen atoms join together because the resulting molecule has less energy than the two separate atoms. The way in which the electrons pair up is a rather more obscure process. Let us imagine an electron moving through a regular two-dimensional array of positively charged ions laid out in a square grid. As the electron travels it exerts an attractive force on these ions and they move very slightly from their idealized positions. A grossly exaggerated case is shown in Figure 10.1. As a result of this disturbance the distribution of positive charges is no longer completely uniform. There is a very small excess of positive charge in the region the electron has just passed through. This extra charge is just sufficient to attract a second electron, and so the two electrons are linked, albeit indirectly. We can picture a much

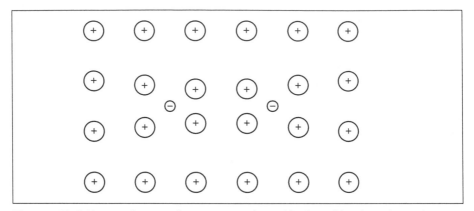

Figure 10.1 If one electron distorts a regular grid of positive ions it produces a concentration of positive charge which attracts a second electron.

simpler demonstration of this effect by considering what happens if a medicine ball is placed on a tautly stretched trampoline. The medicine ball produces a slight depression in the surface, so that if we place a second medicine ball on the trampoline it will tend to roll into the same region.

Although there appears to be no sure way of predicting whether a particular material will form a superconductor, the theory does go some way to relating the superconducting properties of a material to its normal electrical resistance. Copper and silver are good conductors at room temperature precisely because their conduction electrons do not interact strongly with the ions. We would, therefore, not expect these materials to form superconductors since the mechanism for forming electron pairs is very weak. In contrast, many materials which are normally poor conductors exhibit a strong interaction between the electrons and the ions, consequently there is a good possibility of electron pairs forming at low temperatures.

The pairing of the electrons also explains the abrupt change from a normal conducting state to a superconducting state. Such a sudden transition is known as a phase change. A familiar example occurs when a solid melts. For instance, we know that at precisely zero Celsius pure ice turns into water. This is because the thermal vibrations of the molecules at this temperature become so great that the molecules can no longer remain bonded to one another. In a similar way, the thermal motion of the atoms at the critical temperature destroys the tenuous link between the paired electrons in a superconductor.

If we accept that the electrons form pairs, it is still not immediately apparent why the material should offer no resistance to the flow of the electron pairs. The reason is connected with the fact that all the pairs of electrons in a given material have exactly the same properties. This is quite a difficult concept to grasp because it is far too tempting to think in terms of the properties of individual electrons.

In order to make this clearer let us consider a more tangible example involving the speeds of cars on a motorway. Suppose that at some instant we are able to determine the speeds of all the cars on a short stretch of road. The cars travelling in one direction are moving at a variety of different speeds, whilst those travelling in the opposite direction have a similar distribution of speeds. We now conceptually arrange the cars into pairs such that each pair consists of two cars moving at the same speed but in opposite directions. If we take the average velocity of any such pair of cars we of course obtain a resultant velocity of zero. Consequently, although the individual cars have a range of different speeds and are moving in different directions, each pair of cars has precisely the same net velocity (i.e. zero). In a similar way, the electrons in a superconductor form pairs in such a way that the properties of each pair are the same as those of all the others. In the absence of a current there is no net movement of electrons, and so all the pairs have an average velocity of zero. If a current flows, then all the pairs have the same finite velocity—they move at precisely the same speed in a common direction. Consequently, we can see that the electron pairs must move together as a collective body. This is quite a different picture from that of electrons travelling through a normal metal. In a metal the electrons move as individual entities with little care or concern for any of the other electrons. As a result of collisions with impurities or vibrating atoms they are deflected into different states, producing the effect we call resistance. However, when the electrons act as a team the only possible cause of resistance is for all the electron pairs to be simultaneously scattered into a different state, an event which is so implausible that all attempts to measure any resistance in a superconductor have failed.

Indeed, many experiments have been performed in which a pulse of electrical current is left to flow around a superconducting loop of wire. In a normal wire the electrons would undergo collisions with the atoms, and in a very brief time the current would disappear altogether, the electrons ending up moving in completely random directions. In a superconductor there are no collisions, so that even after a duration of many years there is no perceptible change in the current. It is hardly surprising that these tests have been dubbed the most boring experiments in the world. Although these measurements can never confirm that the resistance of a superconductor is precisely zero (to do so they would have to run for an infinite amount of time), they do allow an upper bound to be placed on the resistance of materials in this state. The results indicate that the resistance is at least a billion billion times smaller than that of copper at room temperature.

*

Now that we have a sufficient understanding of how the electrons behave in a superconductor, let us look at how we can make use of this phenomenon in high-speed digital circuits. We will first turn our attention to the interconnects, the tiny strips of metal which form the electrical connections between the individual transistors.

As we have seen in Chapter 5, the electrical resistance is the cause of most of the problems associated with interconnects. In particular, as the dimensions of the interconnects are scaled down, the resistance adversely affects the time taken for signals to propagate between the devices, and in some cases can lead to electromigration, a complete breakdown of the fabric of the wire. The use of zero resistance superconducting interconnects can certainly help to reduce the delay time, and in addition high-frequency signals suffer less distortion than they do in conventional wires. What about the problem of electromigration? This occurs due to the enormous heating effects when the current density (the amount of current divided by the cross-sectional area of the wire) becomes too large. If a superconductor really has zero resistance, then surely heating effects should be absent no matter how large the current density. Unfortunately, it is not quite so simple. To see why let us allow ourselves to be side-tracked for a moment.

It is known that a moving electrical charge produces a magnetic field. This is the principle of operation of the electromagnet: an electric current flowing through a coiled wire is the source of the magnetic field. As we have just seen, the heat produced by a current flowing in an ordinary metal wire places a severe restriction on the amount of current that the wire can carry. Consequently, the magnetic field produced by this effect is quite small. To make a powerful electromagnet it is necessary to wind the wire around an iron cylinder which acts as an amplifier, magnifying the magnetic field produced in the wire. A major disadvantage of this arrangement is that in a large electromagnet the solid iron core is extremely heavy. Onnes' discovery of superconductivity seemed to promise a much better solution. He dreamed of superconducting wires capable of carrying huge currents without resistance, and so producing immense magnetic fields without the burden of a heavy iron core. However, in practice the materials did not live up to expectations. Disappointingly, it was found that the superconductor reverted to behaving as an ordinary metal when anything but the smallest current was passed through it. Ironically, it turned out that it was the magnetic field produced by the current which destroyed the superconductivity! We can understand this behaviour using the Bardeen, Cooper and Schrieffer (BCS) theory of superconductivity by saying that the magnetic field destroys the coupling between the paired electrons. Fortunately, many materials have since been discovered which have a much greater tolerance to a magnetic field. In some cases superconducting wires can carry currents which are thousands of times larger than those that can be handled by ordinary wires of similar dimensions. Indeed, Onnes' vision has been fulfilled. It is now possible to build superconducting electromagnets which are far more powerful than, and only a fraction of the weight of, conventional electromagnets. To date this is the main commercial application of superconducting materials.

Although the effect of a magnetic field in destroying the superconductivity of a material is generally a hindrance, it can be used to advantage in a slightly different situation to perform a switching function analogous to the semiconductor transistor. The crucial step in this direction was made in 1962 by a twenty-two-year-old research student at Cambridge University, Brian Josephson. Before we examine Josephson's results it is helpful to have a bit of background.

A couple of years earlier Ivar Giaever, working at General Electric Laboratories had examined the properties of a thin layer of insulating material sandwiched between two superconductors. Such a system can now be synthesized quite simply. If we allow the surface of a superconductor to oxidize, the oxide forms a suitable insulating layer, and a further layer of superconductor can then be deposited on top. How does the oxide layer affect the properties of the material? In particular, what happens to the electrons when they reach this barrier? We might expect that they will tunnel through the barrier without any problem. However, the situation is rather different from the one we have previously encountered since we are now dealing with superconducting pairs of electrons. Let us suppose for a moment than an electron from one of the pairs tunnels through the barrier. We are now left with an unpaired electron on each side of the barrier. Since we have said that two single electrons reduce their energy by forming a pair, a process which converts an electron pair into two unpaired electrons must increase the energy of the system. We therefore predict that it is unfavourable for an electron to tunnel through the barrier, and so a current does not flow. Suppose that we now increase the voltage very slightly on one side of the barrier so that the electrons on this side are raised in energy. Does this enable tunnelling to take place? That depends upon the magnitude of the applied voltage. Remember that the pairs of electrons have a lower energy than the individual electrons, and so there is a small energy gap separating the quantum state occupied by the electron pairs from the lowest single electron state. If the electron pairs on one side of the barrier have an energy corresponding to the gap on the other side, then there is still no current flow because there are no states into which the electrons can tunnel. However, a very small increase in voltage is sufficient to cause a sudden surge in current as the paired electron states become aligned with the vacant single electron states on the other side of the barrier.

Such a nonlinear relationship between current and voltage suggests that a structure of this type could form the basis of a switching element. Josephson's discovery added an interesting extension to this idea. He predicted that if the thickness of the insulating layer is reduced to about one nanometre (a millionth of a millimetre), it would be possible for pairs of electrons to tunnel through the barrier. In other words, under these circumstances the insulator behaves as though it too is a superconductor! For this work Josephson and Giaever received the Nobel prize in 1973, which they shared with Leo Esaki, the inventor of the tunnel diode. This device, now known a a Josephson junction, exhibits quite remarkable characteristics. The junction region is called a weak superconductor. This is rather like having a chain which possesses a weak link. The strength of the chain is determined not by the properties of the many sound links, but by the strength of the single weak link. This may seem like a design fault, but in fact it can be used to our advantage. By deliberately making one link weaker than the rest, we know precisely where the chain will give way, and so it is necessary only to monitor this one link to determine whether the chain is in danger of being overloaded. In a similar way, by knowing where the superconductor will fail we can use the junction to determine whether or not the device superconducts.

One way to implement this effect is to use the current flowing through the device to control the state of the device. A small current can flow through the insulating layer without any resistance. However, a slight increase in the current produces a magnetic field sufficient to destroy the coupling of the electrons in the junction region. Although single electrons can continue to tunnel through the junction, they now encounter a substantial resistance.

A more satisfactory way to achieve the same effect is to use a separate control current. In many respects this bears a close similarity to a conventional metal oxide semiconductor transistor. A comparatively thick layer of insulating oxide is grown above the junction region and a control wire is placed on top of this layer. This arrangement is illustrated in Figure 10.2. We can now implement this device in a configuration resembling that used to incorporate MOSFETs into a logic circuit. A source of current and an output lead are connected to one side of the junction, whilst the other side is earthed. If we use a current which is slightly smaller than the critical value needed to make the junction fail, then the device will superconduct and the input current flows straight to earth. However, if a small current is passed through the control line, the resulting magnetic field destroys the superconducting behaviour of the junction. Since the device now exhibits a large resistance, the current flows instead through the output wire.

Thus although the physics is radically different, the behaviour is very similar to that of a conventional transistor. The main advantages of a Josephson junction transistor are that it is extremely fast and consumes very little power. The power consumption is of major significance. Since the operating voltage is typically only about a thousandth of that used in conventional microelectronics, and the current is also reduced, this means that the amount of heat produced by a Josephson junction circuit is only about one ten thousandth of that generated by a comparable semiconductor circuit. This means not only that devices can

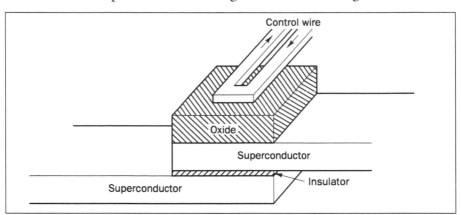

Figure 10.2 A Josephson junction.

be packed together more tightly on a chip, but also that the chips themselves can be placed in much closer proximity as there are no heat dissipation problems to contend with.

The similarity between Josephson transistors and MOSFETs suggests that the circuits can be laid out along very similar lines. In fact, Josephson transistors are slightly more adaptable in that logic gates can be constructed using single devices. As a simple example let us consider an OR gate which produces an output current if either or both of the inputs are present. To perform this function we simply used two control wires. If there is an input to either wire the junction ceases to be superconducting and the current flows through the output channel. The only way in which the output can flow to earth is if both of the inputs are zero.

However, there is one major difference between Josephson junctions and semiconductor devices. A semiconductor device responds in a particular way to an electrical signal, but it returns to its initial state once the signal is removed. In contrast, a Josephson junction remains in the switched state. To see why this occurs let us consider a Josephson junction which is initially in the superconducting state. This means that there is no difference in voltage between the two sides of the junction. Pairs of electrons on one side of the barrier tunnel into similar energy states on the other side. If a current is now applied to the control line so that the junction changes from superconducting to normal mode, then the device develops a resistance and a voltage appears across the junction. As we have seen, this voltage raises the energy levels on one side of the junction so that the pairs of electrons tunnel into single electron states on the other side. If the control current is now removed, the voltage difference still exists and so the system does not return to a superconducting state. The only way we can force

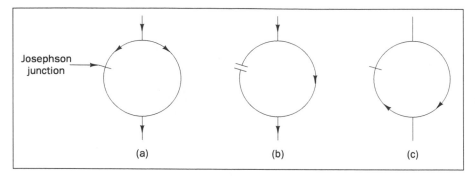

Figure 10.3 Principle of a memory element based on a loop of wire containing a Josephson junction. In (a) the junction superconducts and current flows through both arms of the loop. If the weak link is broken (b), current flows through only one arm. When the input current is switched off (c), a circulating current remains in the loop.

the system back to its original state is to turn off the supply current momentarily. When we do this the voltage difference disappears, so that when the current is switched back on, the junction is once again superconducting.

The need to switch the power off between each operation may seem like a major constraint, but in fact many electrical systems rely on a similar process. Our domestic electricity supply, for example, switches on and off one hundred times a second. Why not use a similar alternating current to power the computer, the only difference being that we require the current to switch several million times faster than the domestic supply? Such an approach could not be used with semiconductor devices, since, as we know from the simple diode, they respond quite differently when the current flow is reversed. However, Josephson junctions are quite indifferent to the direction of current flow, so using an alternating current does not produce any new problems.

We have seen that logic circuits based on superconducting technology compare very favourably with existing semiconductor approaches. The other vital ingredient for a computer is memory. In principle, it is quite straightforward to construct a memory element using a Josephson junction. Suppose that we take a length of superconducting wire and bend it to form a loop. If the ends of the wire are lightly oxidized then our loop will contain a Josephson junction, a weak link, at the point where the two ends join. We now attach leads to the top and bottom of this loop so that the left arm of the loop, say, contains the Josephson junction, and the right arm is a continuous length of superconductor, as shown in Figure 10.3. In this way we have produced an asymmetrical structure in which the right arm is stronger than the left.

To see how this arrangement works, consider what happens when we apply an electrical current to this system via the top wire. On reaching the loop, the current splits into two and passes through both halves of the loop in the same way that a river divides to flow around both sides of an island. Assuming that the current is small enough for the Josephson junction to remain superconducting, the two currents recombine and flow out through the lower lead (see Figure 10.3(a)). If the supply current is now turned off, the system returns to its original state. We do not appear to have achieved very much. However, the scenario is quite different if we simultaneously pass a small current through the control wire above the Josephson junction. The junction ceases to superconduct and we find that all the supply current takes the easier route and flows through the superconducting right arm of the loop. This is shown in Figure 10.3(b). What happens if we now turn off both the source and control currents? Since there is no current flowing through the left arm, there is no voltage across the junction and so it returns to the superconducting state. However, there is no longer any current flowing in at the top of the loop. The only current in the system is flowing in a clockwise direction through the right arm of the loop. As the left arm now also has zero resistance, the current will continue to circle around the loop ad infinitum (Figure 10.3(c)). So depending upon whether or not we apply a current through the control wire, we end up with quite different final conditions. We can interpret the presence of a circulating current as a '1' and the absence as a '0'. This is just what we require from a memory element. There is also the added advantage that the system does not require a continuous supply of power (although it is, of course, necessary to ensure that the temperature is maintained at a sufficiently low level for the wire to remain superconducting).

We have demonstrated how a binary digit can be written to a superconducting memory element. The process of reading the contents of the memory, without altering the state of the memory element, follows very similar principles. Again we use a Josephson junction, this time as part of a separate wire buried underneath the superconducting loop. This second Josephson junction will change state depending on whether or not there is a current circulating in the loop, and is therefore able to determine the contents of the memory element non-destructively.

Although this all sounds very promising on paper, there are many problems in practice. In particular, if the memory elements are packed too tightly together there is a strong chance that the control signal to one element will affect the Josephson junctions in neighbouring elements. Such problems have proved to be the main stumbling-block in the development of a computer based on Josephson junction devices. Many of the advances in superconducting electronics were made as a direct result of a major commitment to this technology by IBM between 1968 and 1983. However, in 1983 IBM abandoned the project due to the problems involved in producing a high-speed memory. Although at the outset the predicted performance of superconducting electronics was far in advance of existing silicon technology, the advances in semiconductor memory over the intervening fifteen years meant that by 1983 there was little to choose between the two. Obviously there is no commercial gain in pursuing a new, and potentially problematic, technology which offers only marginal advantages over a mature technology.

Does this mean that there is no future for a superconducting computer? Far from it. The most promising approach seems to be to combine the best features of both superconducting and semiconducting technologies. In this way a supercomputer could be constructed using a conventional semiconductor memory with the high-speed capabilities of Josephson junctions exploited to their full potential in the logic circuits. It may even be possible to fabricate the two on a single chip since similar processing steps are used in both cases. In fact, most superconducting circuits are at present grown on top of a standard silicon wafer. This is not to take advantage of the semiconductor properties of silicon, but merely because of the high quality of these wafers and the depth of experience gained from many years of working with them.

Josephson junctions have also found applications in other areas. Their extraordinary sensitivity to changes in magnetic fields is used in the delightfully named SQUID, an acronym for superconducting quantum interference device. These devices are used, for example, in magnetoencephalography to monitor the tiny magnetic signals generated in the brain. This requires the capability to detect magnetic fields over a billion times weaker than that of the Earth.

The Josephson junction is also used in metrology as a voltage standard. In this case a slightly different property of superconductors is exploited. We have said that the resistance of a superconductor is zero, or as near as makes no difference, but this is true only if we are talking about direct currents. If the current alternates then this can destroy the coupling of the paired electrons and so give rise to resistance, which in turn means that there will be a voltage difference between two points on the wire. However, very high frequencies are required to produce any significant effects. For example, a current with a frequency of a hundred billion cycles per second produces about one thousandth of a volt across a Josephson junction. Why is this useful in metrology? The reason is that the voltage produced is independent of the characteristics of the junction and of the magnitude of the current. It is related only to the frequency of the alternating current and to the same two fundamental constants as we met in the quantum Hall effect, the charge on an electron and Planck's constant.

*

Having explored some of the properties of superconductors we now take a slight change of direction, although we will still be concerned with metals at low temperatures. We have seen already in this book several instances of how our understanding of the physical properties of solids is still very much in a developing stage. The chance discoveries in the last decade of the quantum Hall effect and high-temperature superconductors have both served to demonstrate that our knowledge of these systems is far from complete, and theorists are still struggling to explain some of the finer subtleties of these phenomena. However, perhaps the most profound recent discovery has come from the study of a seemingly simple scenario of electrons tunnelling through a thin layer of insulator between two small electrodes. A series of calculations and experiments on this system has given us a new insight into the behaviour of electrical currents in solids, produced an entirely new concept in transistor-like devices, and even challenged one of the fundamental assumptions of quantum theory. The remainder of this chapter examines these remarkable findings.

Consider a capacitor consisting of two metal plates separated by an insulator. If a negative charge is applied to one of the plates, then an equal amount of positive charge will accumulate on the other plate. Now let us ask a seemingly innocent question. Is the amount of charge on the plates quantized? In other words, does the amount of charge always increase or decrease by certain fixed amounts?

Think about this for a moment. We know that the smallest unit of isolated charge is the amount contained on a single electron. Obviously, then, the electrical charge on an isolated object is quantized since we can only add or take away whole numbers of electrons. But what about the charge on the surface of a capacitor? We would probably expect that this is also quantized—but we would be wrong!

Let us examine the situation more carefully. We have to remember that the metal plate is composed of both negatively charged electrons and positively charged ions. Consequently, when we talk about the charge on a metal surface we really mean the net charge. Strangely enough, this turns out to be a continuously varying quantity. We demonstrate this in Figure 10.4. In (a) the average charge at the surface is zero since the negative and positive charges cancel each other. A split second later, in (b), one of the electrons has moved a little way towards the surface, while the rest of the electrons have remained fixed. Obviously, the surface is now slightly negatively charged, but we cannot say that the surface has a charge of one electron. This means that the charge associated with the surface must be some fractional amount of an electron charge. In fact, since the electron can move by any arbitrary amount, the charge on the surface must also be able to vary by any amount.

Once again we have been misled by intuition. Stranger still is the fact that until 1985 no-one really gave the matter any thought. However, the fact that a surface can have a fractional amount of an electron charge leads to some very interesting results. We might ask, for example, how it affects our understanding of electrical current. A current arises as a result of discrete particles—electrons—moving through a material. This would tend to suggest that current is quantized, and

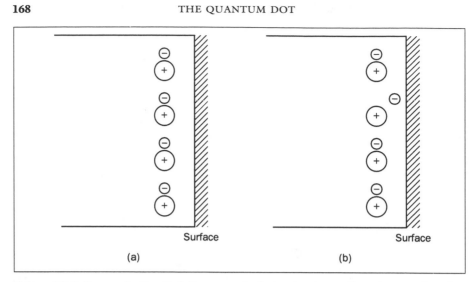

Figure 10.4 Demonstration that the amount of charge at a surface is a continually varying quantity.

yet since the current can be thought of as the movement of a surface of charge through a material we find that it too is a continuously varying quantity. However, this is not always the case.

Consider a structure consisting of two small electrodes separated by a thin layer of insulator. If the system is connected to a constant current source we expect the electrons to tunnel through the insulator. Tunnelling is obviously a discrete process since it must always involve the transfer of a whole electron from one side to the other. We therefore have a very intriguing situation at the interface since the amount of charge changes both continuously and by discrete amounts! In order to resolve this seemingly contradictory state of affairs we must examine the system in more detail.

We will assume that the electrodes are about a tenth of a micron across. Such structures are large enough to be fabricated, but should be small enough so that effects due to the quantization of charge are noticeable. In addition, we require the temperature to be typically within about one degree of absolute zero. As we mentioned when discussing the quantum Hall effect, such a low temperature ensures that the system is always in the lowest possible energy configuration. In this context, the energy of the system is determined by the difference in the charges on the opposing surfaces of the two electrodes.

Let us suppose that the system is initially in the state shown in Figure 10.5(a), where there is no charge on the surfaces of the electrodes. What is the likelihood of an electron tunnelling across the junction, say from left to right? We can see the result of this in Figure 10.5(b). The transfer of an extra electron to the right electrode produces an overall negative charge on this side, which we will write

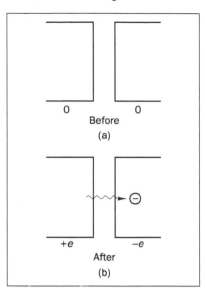

0   0
Before
(a)

+e   −e
After
(b)

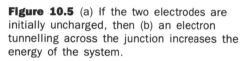

Figure 10.5 (a) If the two electrodes are initially uncharged, then (b) an electron tunnelling across the junction increases the energy of the system.

as −e. Similarly, since we now have one less negative charge on the left-hand side, this will leave behind a net positive charge, +e. The difference in charge across the junction is now 2e, whereas in (a) there was no difference in charge. This means that the process of an electron tunnelling across the junction has increased the energy of the system. From what we said above about the system being in the lowest energy state we must conclude that such an event is forbidden. A similar result is obtained if we consider the probability of an electron tunnelling in the opposite direction. We therefore predict that there is no movement of electrons across the junction. This restriction on tunnelling is called the Coulomb blockade.

How can we lift this blockade? The simplest answer is just to wait. Although there are no electrons moving across the insulator, this does not mean that there is no current flowing in the system. The current will cause a build-up of negative charge on the surface of the left electrode, say, and a similar quantity of positive charge on the right electrode. We soon reach the situation shown in Figure 10.6(a) where there is a charge of half an electron on each side, that is −e/2 on the left-hand side and +e/2 on the right-hand side. If an electron now tunnels across from left to right, as in Figure 10.6(b), it transfers a charge of −e, resulting in a total charge of −e/2 on the right side and +e/2 on the left. Since the difference in charge in each case is equal to e, the energy is the same both before and after the event, making it now favourable for an electron to tunnel across the junction. However, we should note that it is only permissible for a single electron to tunnel from left to right. If two electrons were to tunnel across this would result in a much higher energy state with +3e/2 on the left and −3e/2 on the right.

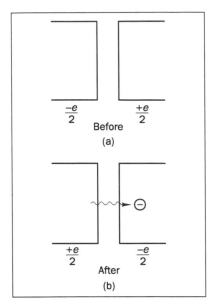

Before

(a)

After

(b)

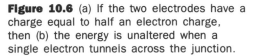

Figure 10.6 (a) If the two electrodes have a charge equal to half an electron charge, then (b) the energy is unaltered when a single electron tunnels across the junction.

From the above discussion we have shown that a single electron is allowed to tunnel across the structure under appropriate circumstances. If we continue to monitor how the system evolves with time we note that further negative charge continues to accumulate on the left-hand electrode. Consequently, the charge on this surface changes from $+e/2$, through zero, and after a short while it is back to $-e/2$ (with an equal and opposite charge on the right electrode). At this point another electron can tunnel from left to right, and the sequence begins again. The flow of electrical current across the thin insulating layer is therefore a series of evenly spaced tunnelling events, each one involving the passage of a single electron across the junction. An elegant analogy for this process is given by Konstantin Likharev, one of the co-discoverers of the phenomenon. We can picture the build-up of charge on the surface of the electrode as being similar to the formation of a water droplet on a leaky tap. The droplet increases until it reaches a critical size, at which point it breaks away from the tap and a new droplet starts to form. The crucial difference is that the droplet of charge which escapes from the electrode is always equal to precisely one electron charge.

A further refinement on the single electron tunnelling structure consists of placing a third small electrode centrally between the two others. If we make an electrical contact to the thin layer of insulator above this electrode, as shown in Figure 10.7, then by altering the voltage applied to the contact we can vary the charge contained on the central electrode. If there is no charge on this electrode, then the Coulomb blockade operates. It is not favourable for an electron to

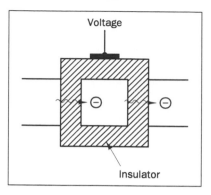

Voltage

Insulator

Figure 10.7 If the central electrode has a charge of +e/2 then electrons can alternately tunnel on to and off the electrode.

tunnel on to the central electrode, since to do so would change the charge on the electrode to −e and so increase the energy of the system. However, if the voltage is used to alter the charge to +e/2, the probability of an electron tunnelling on to the electrode increases dramatically. This is because the addition of an electron changes the charge to −e/2, and since the magnitude of the charge is unchanged, the energy also remains the same. However, only a single electron is permitted to tunnel on to the electrode, since the addition of further electrons would increase the energy. Consequently, one electron must tunnel off the electrode, changing the charge back to +e/2, before another electron can tunnel on to it. This process is known as correlated tunnelling. As one electron leaves the electrode and tunnels through one junction, another electron tunnels through the other junction on to the electrode more or less simultaneously. The process is illustrated in Figure 10.7. We can see that the central electrode acts rather like the gate of a MOSFET, since the flow of current through the structure is determined by the voltage applied to this electrode. As only a tiny voltage is required to switch between the two states, this single electron transistor operates at very low power, and due to the small physical size of the device it is very fast and compact.

A similar structure may also be used in a slightly different context to provide a quantum standard for current. In this case a high-frequency alternating voltage is applied between the electrodes, the voltage between the central and right-hand electrodes being half a cycle out of phase with that applied between the left and central electrodes. In the first half of the cycle the charge on the central electrode is altered to allow just one electron to tunnel across from the left electrode. However, due to the phase difference in the voltages, the electron is not permitted to tunnel on to the right electrode, and so it simply stays on the central electrode, blocking any further electrons from tunnelling. In the second half of the cycle the voltage changes so that it is attractive for one electron to tunnel from the central electrode towards the right, but now it is no longer possible for an electron to tunnel from the left electrode. In this way the system acts like a turnstile for electrons. With each cycle of the voltage a single electron passes through the system. In this way we know the precise rate at which electrons pass through the structure. This means that the current is simply given by the product of the electron charge and the frequency of the voltage.

Finally, let us consider how the size of the third electrode affects the results. We have assumed dimensions of about one tenth of a micron. This is about two orders of magnitude larger than the wavelength of an electron in a metal, and so the effects we have discussed are due only to the quantization of charge. However, we have seen that a region of this size in a semiconductor is sufficient to produce substantial quantum size effects. Consequently, if a similar structure is fabricated from a semiconductor, the central electrode behaves as a quantum dot. This raises an interesting question. Does the single electron tunnelling phenomenon occur when the energies of the electrons are also quantized? The results of experiments on these systems suggest that it does. Exploiting this effect may open up even more possibilities.

Once again we have found that a seemingly straightforward system gives rise to a vast range of unexpected phenomena. In particular, the coexistence of both size and charge quantization has raised many new questions which are still waiting to be answered.

11

Making Light Work
Computing with Photons

IN the last few chapters we have examined many different approaches to producing basic elements for a future microelectronic technology. The one feature which all of these devices have in common is that they modulate an electrical signal. Thus, a current of electrons may or may not flow through the device depending on whether it is in the ON or OFF state. Of course, an ordinary transistor, and even a vacuum tube, performs a similar function. Given this evidence we might well conclude that there are no viable alternatives to electrons. However, during the past two decades an increasing number of researchers have questioned whether electrons are best suited to this task. In consequence, they have pursued an alternative approach in which beams of light are used in the place of currents of electrons. The range and variety of different approaches to the goal of building an optical computer are so broad that we could not hope to cover all of the possibilities in a single chapter. Instead I have chosen to concentrate on those directions which are closest to the electronic transistor.

Let us begin by looking at some of the properties of photons in a beam of light. The replacement of electrons with photons is not an obvious one since the two behave very differently. In particular, electrons are charged particles and therefore interact strongly with one another. As we saw in Chapter 1, electrons also obey the Pauli exclusion principle which says that no more than two electrons can occupy the same quantum state. Virtually every phenomenon which we have described so far in this book is a direct result of either one or both of these properties of the electron. In comparison, photons generally do not interact with one another and are not subject to the exclusion principle. This suggests that any attempt to construct the basic elements of an optical computer will require a very different approach from that which has influenced the development of electronic computers.

The virtual absence of an interaction between photons is very useful when transmitting signals. This has already been exploited in using optical fibres to transmit information across long distances. Many signals can be sent simultaneously along a single optical fibre the width of a human hair, and the fibres can be packed tightly together. This should be contrasted with the use of conventional copper wires, in which the problems associated with interacting electrons are always present. Consequently, the copper wires must be insulated from each other and also from any external electrical signals such as those produced by power lines.

This raises an interesting question. Since light beams have proved so successful in long-distance communication, could they also be used to solve the communication problems within an integrated circuit? Let us consider the advantages of such a scheme. Firstly, electrical signals travelling through metal wires on the surface of the chip can easily interact and lead to 'crosstalk' if the spacing between the interconnects is reduced too far, or if the interconnects are allowed to cross. As a result the routeing of the interconnects must be very carefully planned, and this is a very time-consuming part of the design stage of a chip. The use of optical signals would greatly alleviate these problems. In addition, there is the question of the time taken for signals to propagate through the interconnects. As we have seen, the delays introduced by redistributing information around the chip are due to resistive and capacitive effects. At best these delays remain the same as the circuit is scaled down in size, leading to the conclusion that the operating speed of an electronic computer will ultimately be determined by the time required to communicate signals between the devices. However, optical signals do not suffer from capacitive or resistive effects. The time required to transmit a signal depends only on the length of the interconnect. Therefore, the use of optical signals suggests that the communication time should scale down in proportion with the device size. Finally, one further point comes from Einstein's theory of relativity, which states that a light beam is the fastest way to communicate information from one point to another. The speed of light travelling through a vacuum is thirty thousand kilometres per second. Such a speed is so great that it is hard to comprehend, and yet on the time-scale of microelectronics, typically a billionth of a second, the distance travelled by a light beam is a mere thirty centimetres. (In fact, since light travels more slowly through a medium than in a vacuum, the distance is likely to be somewhat less.) This places a fundamental restriction on the overall size of any high-speed computer. However, it is significant that the speed of light is approximately a hundred times greater than the speed at which an electrical signal travels through a typical interconnect.

The replacement of electrical interconnects with optical ones therefore seems to be an ideal starting-point for the development of an optical computer. Unfortunately, this is not a practical proposition, at least with regard to replacing all the interconnects. The problem of course is that all of the switching elements that we have considered so far require an electrical signal as the input and produce an electrical output. In order to transmit optical signals between devices we would need to construct a laser and a detector to accompany each device in order to do the necessary conversions between electrical and optical signals. This would not only be very costly, but the additional space required would greatly reduce the number of transistors that can be placed on a single chip, which rather defeats the object of the exercise. This does not mean that optical signals will not be used. There is great scope for using optical signals to communicate between different chips, and even for some long interconnections across the chip, but they are not the panacea that they may at first seem.

The problem, of course, is that we are still considering electronic processing elements. Our intention in this chapter was to pursue the possibility of making optical processing elements. This would remove the need to convert between electrical and optical signals, making optical interconnects a far more practical proposition. However, as we shall see, the interconnection needs in these cases are likely to be very different from those in an electronic circuit.

Before we go any further we need to clarify what we mean by an optical processing element. As a general description we can consider this to be a device which accepts an input beam of light, performs some function on it, and outputs the modified beam. Surprisingly, there is a huge range of systems already in existence which can perform such a task. One of the simplest is a single convex lens, the type you would find in a camera, a slide projector or a microscope. In normal use an object is placed on one side of the lens, and produces an image on the other side. The purpose of the lens is to focus the image on to a film, a screen, or the eye of the observer. However, if the object is placed at the focal point of the lens and is illuminated with coherent light from a laser, then the image at the focal point on the other side of the lens bears a special relation to the object. This relationship is described by a mathematical function known as a Fourier transform. We do not need to know exactly what a Fourier transform represents; suffice it to say that it is an extremely useful function which appears in an enormously diverse range of scientific and engineering problems. Consequently, a great deal of computer time is currently employed in calculating Fourier transforms. Let us take a moment to examine the way in which a digital computer performs this task—in doing so we will obtain a vivid demonstration of the enormous potential of optical computing.

Suppose we wish to compute the Fourier transform of a two-dimensional object. In order to perform such a calculation using digital methods we need conceptually to divide the object into many small regions known as pixels. We might typically choose a regular grid measuring a thousand pixels in each direction, giving a total of one million pixels. The computer then has to calculate how each pixel affects all of the other pixels in order to generate the final image. Since every pixel in the object contributes something to each pixel in the image, we would expect the size of the calculation to be dependent on the number of pixels squared, requiring of the order of a million million operations. In fact, a much faster algorithm, known as a fast Fourier transform, has been devised which in this case would take about a hundred million operations. Despite this ten thousandfold saving, it is still an immense task to compute the Fourier transform—it requires at least a few seconds of processing time even on a powerful supercomputer. In contrast, the lens produces the Fourier transformed image in the time it takes a beam of light to pass through the system, typically less than a billionth of a second. How can the lens perform such an amazing feat? It does so because it deals with every pixel simultaneously. To use the jargon we would say that the lens processes the pixels in parallel. Similarly, digital electronic computers can be constructed consisting of several processors. Each of these processors works on a separate part of the problem, and so the final answer is produced in less time. However, the degree of parallelism exhibited by the lens would be unimaginable in an electronic system.

There are other obvious differences between an electronic computer and a lens. Firstly, the lens performs analogue operations. This helps to make it extremely fast, but also makes it less accurate than a digital system. The lens must be very precisely manufactured in order to give a good representation of the Fourier transform, and even then errors will be present in the image due to imperfections in the lens. The other significant difference is that the lens performs a single specific function. It is not possible to program it to perform any general operation, as can be done with the electronic computer. This raises some important issues. For example, should a simple lens qualify as an optical computer, even though it is not programmable? This is a matter for debate. However, we are in danger of straying too far from the theme of this book. Instead we will concentrate on the area of optical computing to which solid-state physics has the most to offer. That is the construction of the basic elements of a general-purpose programmable optical computer.

What do we require of such a device? The most basic requirement is the ability to act as a switch. To see how we might achieve this let us first consider a more familiar example. The electronic transistor acts as a switch because it has two possible output states. These are characterized by the ability of the device to conduct or not to conduct electricity depending on the magnitude of another electrical signal. A similar function could perhaps be obtained in an optical context by devising a device which either absorbs or does not absorb light depending on the intensity of the light beam. However, the lack of interaction between photons makes this exceedingly difficult to achieve. In most materials light passes through without affecting the optical properties of the material. In other words, the passage of one light beam through the material does not affect the propagation of any other beams. An essential requirement is therefore that the presence of one light beam should affect the way in which the material responds to other light beams, and even to the beam which caused the change in the first place. Such materials are said to exhibit a non-linear optical response. A familiar example is the glass used in photochromic sunglasses. Under moderate light conditions the glass appears to be clear, but when subjected to intense light the glass darkens. Thus, a photochromic glass could, in principle, form the basis of an optical switching device, but the speed at which the glass changes from one state to another is much too slow for any computing applications. Typically the glass takes seconds, or even minutes, to darken. In order to construct an optical computer we need a material which changes state in a billionth of a second or less.

Before we consider non-linear optical materials in more detail, let us first take a look at how we could construct a device which exploits this effect. In many cases the device is based on a structure called a Fabry–Perot interferometer. This instrument was devised a century ago by two Frenchmen, Charles Fabry and Albert Perot, who used it to measure the wavelengths of various colours of light. In its simplest form it involves a sample of non-opaque material, two opposing faces of which are cleaved and polished to form mirrors. (In this respect it bears some similarity to the laser diode.) The region between the two mirrors is referred to as the cavity. Let us consider what happens when we shine laser light on to one of the mirrored faces of the sample. Some of the light will obviously be reflected, but since the mirror is not perfect a certain proportion will enter the cavity. This light travels through the sample until it reaches the mirror at the other end. At this point some of the light escapes, but most is reflected back into the cavity. This reflected beam now interferes with the light waves travelling in the opposite direction. The important consideration is to determine whether the interference is constructive or destructive. This is quite straightforward as we have already examined a similar case when considering the waves on a stretched string in Chapter 7. Using these results we can see immediately that if the length of the cavity is equal to an exact number of half wavelengths, the waves travelling in opposite directions interfere constructively. This means that a standing wave forms in the cavity, as shown in Figure 11.1. The incoming light constantly adds to this wave, and at the same time the wave leaks out through the opposite mirror. We can consider this to be the output of the device. On the other hand, if the cavity length is not exactly an integral number of half wavelengths, then the incoming and reflected waves interfere destructively. In this case there is virtually no wave inside the cavity—just as we can not have a standing wave on a stretched string in similar circumstances—and so there is no output. We can therefore see that the relationship between the length of the cavity and the wavelength of the light is crucial in determining whether or not an output is produced.

How do we make such a system into an optical switch? One possibility is to change the frequency, and therefore the wavelength, of the light. In this way it is possible to switch between the destructive and constructive modes. This is fine for a single device, but in order to produce a computer we want the devices to be 'cascadable'. In other words, the output of one device should form the

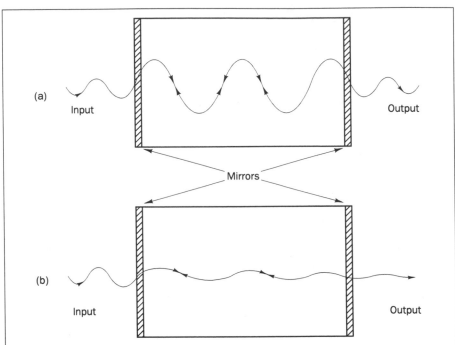

Figure 11.1 (a) Constructive interference in the cavity of a Fabry–Perot interferometer produces a substantial output signal. (b) Destructive interference produces virtually no output signal.

input of another. This cannot happen unless all the devices have the same input and output frequencies. The way to achieve this is to replace the material in the cavity with a non-linear material, typically one whose properties depend on the intensity of the light. How does this help?

To begin with we need to understand the property of refraction, which describes how light interacts with a given material. In particular, refraction is associated with the speed at which light travels through the material. Light travels fastest in a vacuum. When it enters a material, the frequency at which the light oscillates stays the same but the wavelength is reduced. Since the speed of light is equal to the product of the wavelength and the frequency, this means that light travels more slowly through a material than it does in a vacuum. A convenient way to measure this effect is the refractive index, which represents the ratio of the speed of light in a vacuum to its speed in a particular solid. The change in refractive index as light travels from one material to another is often observed as a bending of the light rays. This gives rise to many familiar phenomena. For example, it accounts for the optical illusion which is created when a straight object, such as a pencil, is placed partly in water. The bending of the light rays gives the appearance that the pencil is bent at the surface of the water.

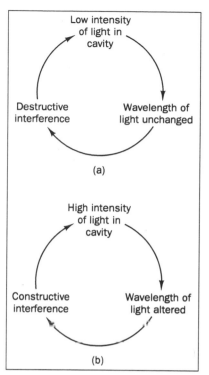

(a)

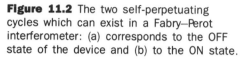

(b)

Figure 11.2 The two self-perpetuating cycles which can exist in a Fabry–Perot interferometer: (a) corresponds to the OFF state of the device and (b) to the ON state.

The refractive index of a material is dependent on the wavelength of the light. As a result different colours of light are deflected by varying amounts. In a prism this gives rise to the splitting of white light into a spectrum of its constituent colours. There is also a less familiar aspect of refraction which is that the refractive index of a materal depends on the intensity of the light. The changes are so small that under most conditions the effect is negligible. It is only with intense beams of laser light that any substantial effect can be observed.

Suppose that we place a non-linear material of this type in the cavity of a Fabry–Perot interferometer. We irradiate the system with light, the wavelength of which corresponds to a condition of destructive interference in the cavity. This means that there is only a very weak beam of light in the cavity. The non-linear material is virtually unaffected by this, so the wavelength of the light in the cavity is relatively unchanged and the condition of destructive interference persists. We appear to be trapped in a vicious circle, as we can see from Figure 11.2(a). Since there is no output in this condition (see Figure 11.1(b)) we will say that the system is in the OFF state. If we now increase the intensity of the incident light, then the intensity, and therefore the wavelength, of the light in the cavity also changes slightly. At some point the change in wavelength is sufficient to move the system to a situation where constructive interference dominates. At this point the intensity in the cavity increases rapidly, causing a further reduction in the wavelength, which in turn brings the system closer to the condition of constructive interference. Within a very short time, typically a few picoseconds (a picosecond is a thousand billionth of a second), the amplitude of the light wave in the cavity increases dramatically and an output light beam is produced. There is therefore a certain critical intensity at which the system switches from the sequence of events shown in Figure 11.2(a) to that in (b). This is a resonant effect, just like the one with electron waves produced in the resonant tunnelling transistor. It is important to note that the only quantity that we have altered is the intensity of the incident light beam. Although the wavelength of the light changes within the cavity, the frequency of the light remains constant at all times. Therefore, when the light beam exits the material the wavelength reverts to that of the input beam.

Consequently, the inclusion of a non-linear material in a Fabry–Perot interferometer allows the device to switch between two different states, the switching condition being dependent on the intensity of the input beam. The other important feature of an electronic transistor is the ability to amplify a signal. Although we have emphasized the switching function, amplification is important in a digital computer since the output of one transistor may be passed to two or more transistors. Is it possible to obtain amplification of an optical signal using the above device? Let us examine the system again. When the device is in a condition of constructive interference, which we will call the ON state, the amplitude of the wave inside the cavity may be many times greater than that of the input beam. However, since only a small portion of the wave leaks out, the output intensity is considerably smaller than this. In fact, if we consider the system to be in a steady state so that the amplitude inside the cavity remains constant, then the intensity of the output beam can not be any greater than the intensity of the input. This seems to suggest that the device does not amplify the signal. However, the situation is reminiscent of a bipolar transistor. In the transistor, amplification is achieved by using a weak current to modulate the supply current which flows into the device. In a similar way we can configure the optical device so that there is a beam of constant intensity which maintains the device just below the switching-point. To cause the device to switch only a weak signal needs to be added to the constant beam in order to push the system into the resonant condition. The device therefore acts as an amplifier: a small change in the input produces a large change in the output.

We can also see that in this configuration a single such device can be used to perform different logic functions by appropriately choosing the intensity of the constant beam. An example is illustrated in Figure 11.3, where it is assumed that there are two input beams. If the intensity of the constant beam corresponds to point B, then if either one of the inputs is TRUE, this is sufficient to increase the intensity to C, and so the device switches on. Alternatively, if a weaker

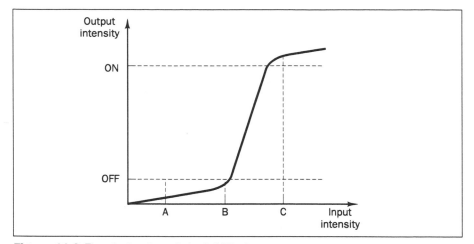

Figure 11.3 The device is switched OFF when the input intensity is at A or B, but switches ON when the intensity is increased to C.

constant beam is used which fixes the intensity at point A, then both inputs must be TRUE for the intensity to reach point C. In the first case the system behaves as an OR gate, while in the second it behaves as an AND gate.

Under certain circumstances the optical system possesses another characteristic which is not present in electronic transistors: bistability. This means that at some values of input intensity there can be two quite different possible outputs. An example of the response of such a system is shown in Figure 11.4. To understand how the system behaves in this way let us consider the following sequence of events. Suppose that the system is initially in the OFF state. In order to switch the device on we have to increase the intensity of the input light beam to the point E. The wavelength of the light in the cavity now corresponds to that required for the resonant condition. The surprising fact is that once the device is in the ON state, it can remain in this condition even if the input intensity is reduced slightly. How do we explain this? The key to understanding this is to realize that the condition of constructive interference occurs if the intensity of the light *within* the cavity exceeds a certain critical value. When the device is switched OFF it is very hard to achieve this critical intensity because the sequence of events in Figure 11.2(a) causes the condition of destructive interference to persist. However, once the device is in the ON state the system responds as in Figure 11.2(b). If we reduce the intensity of the incident light beam, the intensity within the cavity also reduces slightly, but it will still be greater than the critical intensity. The condition of constructive interference is therefore self-perpetuating. It is only if the input intensity is reduced significantly, to the point D, that the system can no longer maintain this condition and so returns to the OFF state. Consequently, at an intermediate value of input power (such

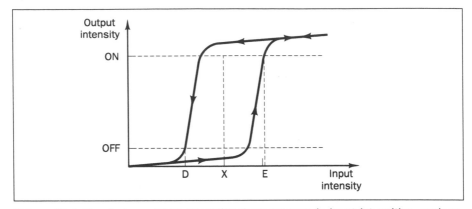

Figure 11.4 If a device exhibits bistability there are certain input intensities, such as X, at which the device may be either ON or OFF depending on the recent history of the system.

as the point X), the device may be either ON or OFF, depending on the recent history of events.

We have seen how a non-linear optical material can transform the behaviour of a device. Let us now look at how this non-linear refractive index is achieved. In particular, what goes on within these materials to make them respond in the way they do? There are several different and complex processes which contribute to this behaviour, but in many cases the main effect is a process called band filling. To illustrate this effect let us consider a semiconductor. We know that if light of the appropriate wavelength is directed on to a semiconductor then electrons can be excited from the valence band into the conduction band. If we choose the wavelength of the light so that the individual photons have an energy just slightly greater than the band gap of the material, then the light is absorbed. Let us suppose that a very intense beam of light is used so that it is possible to excite virtually all the electrons from the highest energy states in the valence band up into the conduction band. As we can see from Figure 11.5, the top of the valence band is virtually empty whilst the bottom of the conduction band is full. This condition is called saturation. Any further transitions would require an electron to be promoted from a lower energy state in the valence band up to one of the higher vacant states in the conduction band. Since this energy change is larger than can be supplied by one of the photons in the light beam, no such processes take place. Consequently, the behaviour of the material alters dramatically. The photons are no longer absorbed and so the material becomes virtually transparent.

Much of the early work on this phenomenon concentrated on the semiconductor indium antimonide, which has a much smaller band gap than materials such as silicon and gallium arsenide. The magnitude of the band gap is particularly relevant since a simple rule-of-thumb states that the number of energy levels at

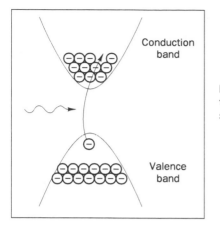

Figure 11.5 When saturation is achieved there are virtually no valence electrons in suitable positions to absorb a photon.

the top of the valence band and the bottom of the conduction band increases with the magnitude of the band gap. Consequently, there are relatively few states near the band edge in indium antimonide, making it much easier to achieve the band filling condition. However, there are also disadvantages to having a small band gap. In particular, since the electrons need only to acquire a relatively small amount of energy to move up to the conduction band, there is a large number of thermally created electrons and holes at room temperature. Since these tend to obscure the effect that we are trying to observe, it is necessary to operate the device at very low temperatures (typically seventy-seven Kelvin, the boiling point of liquid nitrogen). An attractive alternative is the use of wider band gap quantum wells or superlattices. In these materials the number of electron states at the band edge can be engineered by tailoring the properties of the system, and the wide band gap means that the device can still operate at room temperature.

Many experimental systems based on the Fabry–Perot interferometer have been demonstrated, notably by Desmond Smith's group at Heriot-Watt University in Edinburgh and by Hyatt Gibbs' group in Arizona. One undesirable characteristic of these systems is the amount of power required in order to excite a large number of electrons into the conduction band. This in turn means that a great deal of heat is generated, which must be dissipated. The problem is a serious one because the switching intensity depends critically on the length of the cavity. If the system heats up and the material expands, then the cavity length changes. Another problem concerns the switching speed. One of the advantages of optical computing which is often cited is that it is much faster than using electronic devices. However, the switching times quoted for these devices should usually be treated with a certain amount of caution. Certainly, the resonator can switch ON much faster than an equivalent electronic transistor. Typically this process is measured in picoseconds (a picosecond is one thousand billionth of a second). However, before the optical device can be switched ON for a second time, it is necessary for most of the conduction electrons to have returned to the valence band. This is a relatively slow process, the recovery time being of the order of a billionth of a second. In a computer we are usually interested in the time taken for the device to perform a complete cycle, ON–OFF–ON, so from this point of view the optical device may not operate much faster than its electronic counterpart.

However, these restrictions occur only because real excitations take place in the material. What we mean by this is that the photons supply the energy to raise electrons into states in the conduction band. Is there any other way in which a beam of light can affect the material? It seems unlikely. For instance, suppose we use a beam of light in which the energy of the photons is somewhat less than the band gap of the material. Since no absorption takes place, the material is transparent at these wavelengths. It would seem then that the material is unaffected by such a light beam. However, this is not quite true. Although the energy of the photons is rather less than that required to excite a valence electron into the conduction band, the electron can make use of Heisenberg's uncertainty principle to borrow the extra amount of energy. Of course, the condition of the loan is that it be repaid in full before the allotted time period expires. We can picture the valence electron absorbing the photon, making the transition to the conduction band by borrowing the extra energy, and then returning to the valence band and re-emitting the photon before the end of the time period. This is called a virtual transition. The typical time scales involved are extremely short, of the order of femtoseconds. This is just a million billionth of a second. To put this in perspective, a beam of light would travel less than one micron in such a time. It would seem that virtual transitions provide the ideal solution for producing ultrafast optical switching. Since no real absorption takes place, the amount of heat produced by such an operation is negligible, and the process is exceedingly fast. A device based on this principle could possibly operate many thousands of times faster than any electronic transistor. The drawback is that in all of the materials which have been examined so far, the effects are too weak to be of use in any practical devices. This poses the challenge of designing new materials, such as superlattices or other quantum-effect structures, engineered in such a way as to optimize this response.

We have concentrated on the use of the Fabry–Perot interferometer, but researchers have explored several other device concepts. One of the most promising also makes use of quantum structures, but before describing how it operates we need to introduce a new particle, called an exciton, and consider how the optical properties of a semiconductor are affected when a voltage is applied.

The exciton consists of two familiar particles, a hole and a conduction electron. Since the electron has a negative charge and the hole has a positive charge, the

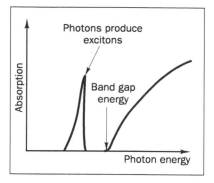

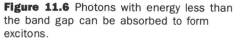

Figure 11.6 Photons with energy less than the band gap can be absorbed to form excitons.

two are attracted together and form a system resembling a hydrogen atom. Just as the components of a hydrogen atom reduce their energy when they bond together, so the exciton has less energy than an isolated hole and conduction electron. We can represent this by picturing the exciton as occupying a state near the top of the band gap, the separation in energy from the conduction band edge being equal to the binding energy of the exciton. Consequently, when the electron and hole in the exciton recombine a photon is emitted which has slightly less energy than the band gap. Similarly, it is possible for a material to absorb photons with an energy just less than the band gap to create an exciton.

The presence of the exciton should show up as a sharp peak in the absorption spectrum at energies just below the band gap, as in Figure 11.6. Such a feature is observed in bulk materials at very low temperatures, but at room temperature the effects are obscured because the difference in energy between the exciton and the conduction band edge is small in comparison to the average thermal energy. However, the effects are much more noticeable when the exciton is confined in a quantum well. This result is simple to understand. Because of the wavelike nature of the electron and the hole, the exciton occupies a certain volume of the crystal, typically with a diameter of about 30 nanometres (a nanometre is a thousandth of a micron). If we confine the exciton within a region smaller than this, then we force the electron and hole into closer proximity. This increases the strength of the bond between them, and so the exciton state is moved down further into the band gap. As a result the sharp absorption peak associated with the exciton is observable even at room temperature. We will use this result in a moment.

Let us now turn our attention to the effect of applying a voltage to a semiconductor. Since the voltage has the effect of tilting the energy bands, a valence electron sees lower energy conduction states available towards the positive end of the crystal. If an electron can make an 'oblique' transition to these states then it is possible for photons of lower energy to be absorbed than when no voltage is applied. This effect, known as electro-absorption, is illustrated in

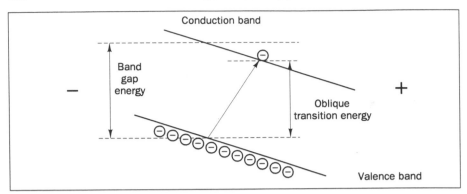

Figure 11.7 Tilting of the bands when an electric field is applied means that oblique transitions may occur with energy less than that of the band gap.

Figure 11.7. However, the difference in energy produced by this effect is grossly exaggerated in the figure. Since the extent over which such an oblique transition can take place is limited to a few tens of nanometres, very large electric fields are needed to produce a significant change in the absorption threshold.

There is also an adverse side-effect because of the way in which the voltage affects an exciton. The electron tends to move towards the positive end of the crystal, while the hole moves towards the negative end. Since the two particles are pulled in opposite directions, this will reduce the binding energy of the exciton, and possibly destroy it altogether. This tends to increase the absorption threshold, and so to some extent counters the electro-absorption effect. However, the problem can be resolved by confining the exciton in a quantum well. In this case we have to picture the electrons and holes as waves. We can see the effect of the electric field on these waves in Figure 11.8. An alternative (classical) way to visualize this is to say that neither particle can move far before coming up against the confining barriers of the well. Consequently, large electric fields can be tolerated without destroying the exciton or even affecting the binding energy significantly.

We now have the required characteristics for building a device which is capable of switching light beams. Let us see how we can achieve this. Suppose we place the quantum well between two thicker layers of semiconductor, one doped p-type and the other n-type. In this way the quantum well forms the central region of a diode. What happens if we shine a beam of light on to this system? That, of course, depends on the wavelength of the light. In this case we want to choose the wavelength very carefully so that the energy of a photon in the light beam corresponds to the energy required to form an exciton. This produces strong absorption, as we can see from Figure 11.9. Having fixed the wavelength of the light, we now apply a reverse biased voltage to the p-n junction (Figure 11.10). We have seen that the effect of a voltage is to shift the exciton absorption

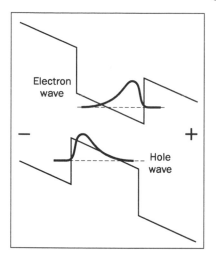

Figure 11.8 The effect of an electric field is to move the electrons and holes in opposite directions, but when confined in a quantum well the carriers cannot move far.

peak to lower energy. (This is true whether the voltage produces a forward or a reverse bias. We will see in a moment why it is important for the junction to be reverse biased in this case.) Consequently, the photon energy now corresponds to the low absorption region of the spectrum, as shown in Figure 11.10. We can therefore vary the absorption by changing the applied voltage. This is a hybrid system, because we are using an electrical signal to control how the system responds to a light beam. Although it may have some uses in interfacing between electrical and optical systems, it is not what we desire for an optical computer. However, the device is rather more sophisticated than it may at first seem.

Let us again consider the system with a reverse bias applied, as in Figure 11.10. We have seen that when it is weakly illuminated with the chosen wavelength of light there is virtually no absorption. We will denote this as the OFF state. We now increase the intensity of the light substantially. How does this affect the response of the system? The argument is simplest if we consider the following sequence of events. Although the energy of the photons corresponds to a region of low absorption, a small proportion of the photons will be absorbed. As we saw in Chapter 3, this generates a voltage, but the voltage will be of opposite polarity to the externally applied reverse bias. In other words, the overall voltage will be reduced slightly, and the exciton absorption energy will increase accordingly, becoming closer to the energy of the incident photons. This in turn means that a greater proportion of the photons will be absorbed, producing a larger photovoltage, and so reducing the overall voltage further still. In general we refer to such a process as feedback. If the intensity of the light is large enough, the sequence continues until the energy of the photons is equal to the exciton peak. Since this produces strong absorption we say that the device is now ON.

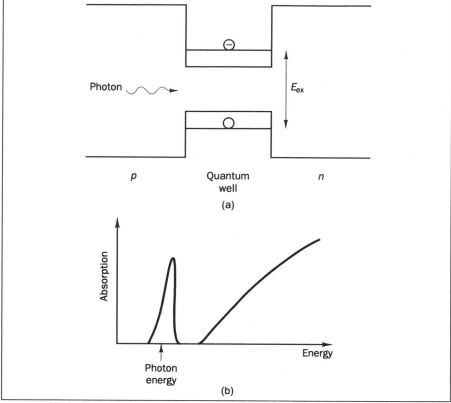

Figure 11.9 (a) The structure described in the text. (b) The photon energy is chosen to coincide with the exciton absorption peak at energy E_{ex}.

We have now reverted to the condition shown in Figure 11.9 where there is no net voltage. However, it is important to realize that we have not actually changed the voltage which is applied to the system. All that we have done is to change the intensity of the incident light beam and the device has done the rest.

The device that we have described is called a SEED, an acronym for self-electro-optic effect device, which refers to the fact that the feedback mechanism is internal to the device. It was developed at AT&T Bell Laboratories in New Jersey by David Miller, a former student from Smith's group at Heriot-Watt. We have seen how a SEED can be used as a switch, responding to the intensity of the incident light. It can also be configured to exhibit bistability, since once the device has reached the high absorption condition it can remain in this state even when the incident intensity is slightly reduced. A SEED therefore performs similar functions to a Fabry–Perot interferometer, but achieves this by a very different physical process. Comparing the two we find that the SEED generally has a lower operating speed, but has the advantages

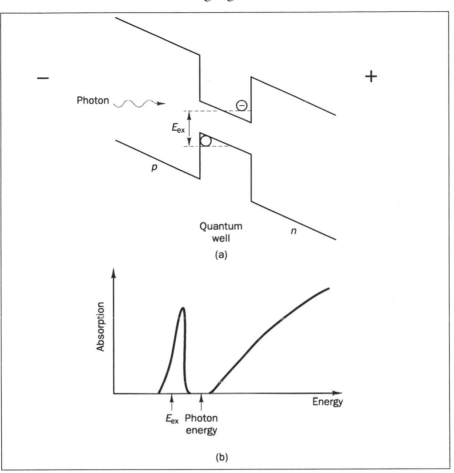

Figure 11.10 The structure shown in Figure 11.9 subject to a reverse bias voltage. The voltage changes the energy of the exciton, E_{ex}, so that the photon energy no longer coincides with the exciton absorption peak (b).

of requiring only low power and, perhaps most significantly, of being much easier to manufacture. The main difficulty in producing an array of Fabry–Perot interferometers is that the wavelength at which each device operates depends critically on the length of the cavity. Although it is a simple matter to tune the wavelength to demonstrate the operation of a single interferometer under laboratory conditions, it will be a much harder task to fabricate a million such devices on a single chip, all of which operate at the same wavelength (in other words requiring that all of the devices have the same cavity length to a very high degree of accuracy).

There is a certain similarity to the competing electronic technologies of MOSFET and bipolar transistors where MOSFETs have become the dominant technology. For similar reasons the SEED may prove to be the more successful device because it is more suited to an integrated optics approach than the resonator, despite the fact that it is significantly slower.

We have explored some of the possibilities of optical computing, concentrating on semiconductor-based optical switches, and have seen how they can be used to create logic gates. However, these devices do not possess all of the characteristics of an electronic transistor, and consequently are not suitable for use as a basis for constructing a digital optical computer. (We will discuss this in more detail in the next chapter.) Nevertheless, the construction of a true optical transistor does not seem to be out of the question and many researchers are currently working towards this goal. Let us suppose for one moment that we have a working optical transistor. How would we implement such a device? One possibility is simply to take a state-of-the-art microelectronic circuit and replace all of the transistors with their optical equivalent, and all of the metal interconnects with waveguides. However, this does not take advantage of the inherent parallelism in optical systems which we have demonstrated with the example of the lens. Future developments are likely to concentrate more on developing arrangements which exploit this parallelism.

Computing the Future

I N the last few chapters we have examined some of the strange effects which occur in the quantum world and seen how these effects can be used in a vast range of novel device concepts. Which of these is most likely to succeed the silicon MOSFET? There is certainly no shortage of contenders for the crown. However, before we address this question let us first see how far we can go with the conventional technology.

In the thirty-five years since the invention of the integrated circuit the number of devices on a single chip has increased from one to about 64 million. To return to the chess board analogy that we introduced at the very beginning of this book, the repeated doubling has taken us to the twenty-sixth square. Most predictions suggest that we will be able to continue scaling the MOSFET down in size to the point where the integrated circuits contain one billion devices—sufficient to move us another four squares along the chess board. What then? Any further reduction in size will take us into the regime where quantum effects become important. We then have two options—either try to make the devices work as before in spite of these problems, or alternatively follow the route that we have pursued throughout this book—exploit the quantum effects in order to design entirely new devices.

Actually, there is also a third option—keep the device size as it is but increase the area of the circuit. An extreme case which is currently being explored is to use the whole area of the wafer to produce one very large integrated circuit, a technique called wafer scale integration. Of course, there will be faults in certain parts of the circuit due to imperfections and contamination, but this may not hinder the operation of the entire chip if a certain amount of redundancy is built in, for example by duplicating the most important circuits. Another possibility is to extend the circuit into the third dimension. We have already seen that in many cases the network of interconnects occupies a series of levels above the surface of the chip. What then is to stop us constructing several layers of devices stacked one above the other, separated from each other by a thick insulating layer? The main difficulty is that of dissipating the heat produced by devices in the centre of the structure, but methods may well be found to circumvent this problem.

However, we should remember that merely increasing the number of devices on a chip is not the foremost concern. The most important criteria are those of cost of fabrication and speed of operation. Furthermore, most of the designs that we have considered do more than simply switching between two states. A single device can be used as a logic gate or as a memory element. To obtain a true comparison we should therefore consider the cost and time of performing a specific function. How do the alternative device concepts compare on this basis?

Speed of operation is the forte of these quantum devices, but to build a fast computer it is not sufficient merely to have a device which can change state rapidly. In order to take advantage of this ability it is necessary to alter the computer architecture, that is to reorganize the arrangement of the units within the computer. This is by no means a new idea. One approach is parallel processing. We have already mentioned this briefly in the previous chapter, but let us now take a more detailed look at the advantages of parallel processing. Until quite recently most computers on the market used sequential processing. This means that the computer possesses a single processor. Each task is broken down into a series of basic operations which are then handled one at a time by the processor. An obvious improvement is to use multiple processors, allowing the computer to work on different parts of a problem at the same time. This is precisely what we mean by parallel processing. Whilst this approach can give quite a significant increase in performance, the main problem is that the different processors must be able to communicate efficiently with one another so that the results generated by one processor can be used by another. As the number of processors increases, so the amount of time that the processors spend communicating with each other also increases. Consequently, using more processors does not necessarily produce a corresponding decrease in the time taken to perform a particular calculation. In addition, the problem of physically connecting each processor to all of the others grows exponentially as the number of processors increases. However, in an optical system the lack of interaction between two or more beams of photons means that these problems do not occur. As we saw in the previous chapter the degree of parallelism exhibited by a simple optical system such as a convex lens is of the order of a million. Consequently, an optical computer possesses an inherent degree of parallelism far in excess of what can be achieved with an electronic system.

Another alternative architecture is cellular automata. In this case the processing unit of the computer is made up of a large number of cells, each containing a relatively small number of devices, which communicate only with neighbouring cells. The main advantage of this strategy is that it eliminates the need for long interconnects between devices which we have seen will ultimately limit the speed of a conventional chip. We examined such a system in Chapter 9 where we considered each cell to consist of five quantum dots. The cells can be arranged to mimic the logic gates of a conventional computer, or in a more adventurous approach arrays of cells could be used to perform highly complex functions. Although these ideas are still at a very early stage of development we can make several forecasts about the properties of quantum dot cellular automata. Firstly, it is already possible to fabricate quantum dot cells of the type described in Chapter 9 which are several times smaller than the predicted minimum dimension of MOSFETs, and with improvements in the fabrication techniques the dimensions of the cells could be scaled down still further. In addition, the cells require only a tiny fraction of the power needed to operate a conventional transistor, and the switching speeds are incredibly fast—predictions suggest that an array of cells will be able to compute the output to a complex function in considerably less time than it takes a single transistor to change state.

Let us now consider the cost associated with the new technologies. This is very difficult to assess since in most cases the devices have only been constructed on a one-off basis for investigation under laboratory conditions. It is an enormous step to take a device from this stage and turn it into a commercially viable technology using mass-production fabrication. However, there have been a couple of notable attempts to dislodge silicon. One major effort which has involved many companies and research establishments worldwide for the last three decades is the introduction of gallium arsenide transistors. These devices operate considerably faster than their silicon counterparts, but many difficulties have been encountered, particularly in the fabrication of the integrated circuits. As a result the level of integration achievable with this material lags some way behind silicon, and for this and other reasons gallium arsenide circuits are far more expensive than silicon ones. The additional problem is that throughout this period of development, silicon technology has continued to evolve rapidly. This has made it very difficult for gallium arsenide technology to become established. Consequently, the main applications have centred around areas where silicon can not compete, such as high-speed analogue circuits (using low levels of integration) and integration with optical systems where the indirect band gap of silicon means that it is unsuitable for these applications. Despite the promise of much higher operating speeds than can be obtained with silicon transistors, it is only within the last few years that gallium arsenide has succeeded in making any inroads into the supercomputer market, with the introduction of the Cray 3 machine.

The other significant attempt to overthrow silicon was a particularly brave effort by a single company. Starting in 1968, IBM launched a massive project to convert the fledgling Josephson junction technology into a viable form suitable for constructing a computer based on superconducting devices. Although several devices based on the Josephson effect had been demonstrated in the laboratory, the task involved was monumental. It required that high-density integrated circuits be mass produced at low cost using what were essentially unknown materials. After fifteen years and an estimated cost of $300 billion the project came to an end. Although enormous progress had been made in this time, one of the principal factors affecting the decision to terminate the project was that in the intervening period silicon technology had advanced to such a stage that the difference in predicted performance of the two systems was down to a factor of only two.

Examining these case histories we can identify two major problems which face any new technology. Firstly, converting prototype devices into high-density circuits which can be mass produced requires enormous amounts of time and money, most probably beyond the means of any single company. The second problem is one of overcoming the inertia of the current silicon technology. Since this established technology is still continuing to develop at a terrific pace, during the time period required to develop any new technology silicon may well have caught up or even surpassed the expectations of the new approach.

It seems, at least in the short term, that future developments favour evolution, rather than revolution. The most likely route for the introduction of new technologies will be through hybridization of the existing technology. For example, it may be possible to incorporate some quantum effect transistors into a conventional circuit to produce a few very fast devices for a specialized purpose within the chip. Although most such devices that we have considered are based on gallium arsenide and aluminium gallium arsenide, the continued dominance of silicon may mean that similar devices using thin layers of silicon and germanium may be of most importance. In addition, problems of communication between devices on a chip may be solved by introducing optical or superconducting interconnects. The most likely application of an entirely new technology to appear in the market place in the near future is the optical computer. By exploiting the massive parallelism available from such a system and applying it to problems such as those encountered in image processing, optical systems would not be in direct competition with the traditional areas of silicon-based computing.

So far in this concluding chapter we have concentrated on the application of these new technologies to computers. However, not all of the devices that we have described are suitable for forming the basic elements of a digital computer. The reason for this is that we need far more than a device which simply switches on and off. The most important feature of a digital computer is that data are not distorted or degraded even after many thousands of operations. In order to achieve this it must be possible to identify the value of a digit with absolute certainty at each stage of the calculation. In addition the devices must exhibit considerable tolerance to the effects of noise. What we mean by this is that in a large circuit all the devices that are in a particular state must produce the same output voltage even though the devices may differ slightly, the temperature may vary from one point to another, and the gate voltage may be slightly different from the expected value. We have seen in Chapter 4 that the failure of Babbage's

Analytical Engine can be attributed to the fact that the tolerance of the individual components was not sufficient to enable the distinction between ten different states. In contrast, the success of the transistor is due to the fact that when configured in a switching circuit the device has two very well-defined states, ON and OFF. It is important to note that it is not just the number of different states which is significant, but also the fact that the output remains constant for a wide range of inputs. The same can not be said of some of the novel devices that we have considered. For example, the resonant tunnelling transistor that we examined in Chapter 9 is extremely sensitive to small changes in the input voltage. The electron interference devices, such as those shown in Figures 9.9 and 9.10, are far less sensitive, but they still do not display well-defined states as the following argument shows. Let us suppose that a voltage of exactly 0.5 Volts is required to switch the device off. This means that a condition of destructive interference occurs—the electron waves cancel each other out and the output is precisely zero. If instead we apply a voltage of 0.51 Volts then the cancellation is no longer complete and there is some residual output. This does not seem to present a significant problem since we can clearly still identify the output state as the state 0. However, the cumulative effect of many such operations will introduce errors into the final result, and so such a device is not suitable for use in a digital computer. Consequently, much of the effort in current research is directed towards those devices which can fulfil the above criteria. These include certain optical devices and the quantum dot cellular automata mentioned previously.

Even though not all of the devices are suitable for digital applications, their high speed of operation means that they could have considerable uses in analogue systems. Many research groups are currently looking into the possibility of building analogue computers in the hope of producing a machine which can perform many of the tasks for which digital computers have proved unsuited, such as shape recognition. Some of these new devices could well form the basis for such a machine. In addition there are many other areas of application outside computing. Resonant tunnelling transistors have already been demonstrated to function as amplifiers dealing with signals which oscillate up to four hundred billion times every second, whilst modulation doped transistors are already commercially available and are used for amplifying high-frequency signals in satellite receivers.

The technology which will have the most significant effect in the near future is likely to be the semiconductor laser. Double heterostructure and quantum well lasers have the capability to be made very small, are highly efficient, and as a consequence require very low operating power. Since they can be readily incorporated into conventional circuits they are also very cheap to produce. Already such devices are in use in compact disc machines, laser printers and in fibre optic communication systems. Specialized chips have also been constructed which contain more than a million tiny lasers. Although these are still at the development stage there is a huge range of potential applications for such a structure. For instance, it could be used to form a tiny but high-resolution television screen or computer monitor small enough to fit into a wristwatch.

Aside from these issues, the phenomena we have discussed are interesting in their own right. We might ask: do particle physicists worry about practical applications of the Higgs boson? Of course not. To paraphrase a statement normally applied to mountaineers, physicists investigate these phenomena because they are there. In doing so we have entered into an entirely new world. Even in the quite recent past, the physical processes taking place in solids were assumed to take place on two length scales, one being the microscopic level represented by individual atoms, and the other being the macroscopic level more akin to our everyday experience. We have now discovered that many interesting effects occur on what has been called the mesoscopic scale, which is somewhere between these two extremes. Some of these phenomena have already found applications. For instance, the Josephson effect is used in SQUIDs to measure minute magnetic fields, and along with the quantum Hall effect has become an important tool in metrology. The future applications of these effects may be far beyond anything we can currently envisage. One thing, however, seems certain. Whatever the future holds, the rich variety of phenomena that occur in the quantum world is sure to be a source of inspiration for many years to come.

GLOSSARY

absolute zero the lowest temperature theoretically possible, corresponding to -273.15 Celsius.

acceptor an impurity atom which has fewer valence electrons than the atom it replaces. When introduced into a semiconductor it produces a hole.

Aharanov-Bohm effect the phase of an electron wave is altered as it passes near to a magnet or an electric charge. The Aharanov-Bohm effect predicts that this occurs even if the magnet (or electric charge) is completely shielded from the electron. This has recently been verified experimentally, but complete shielding is not necessary for the applications considered in this book.

band gap the energy difference between the conduction and valence bands, and equal in magnitude to the energy required to create an electron–hole pair.

base the central region of a bipolar transistor.

BCS theory the theory which describes the behaviour of electrons in a superconductor. It is named after Bardeen, Cooper and Schrieffer, the co-authors of the theory.

binding energy the amount of energy required in order to remove a particle from a given system. For example, the binding energy of the electron in a hydrogen atom is given by the energy required to remove the electron far from the nucleus.

bipolar transistor one of the two main types of transistor. The regions of the device are known as the emitter, base and collector. Depending on the type of doping present in each of these regions the transistor is described either as a pnp or an npn transistor.

CMOS a particularly favourable arrangement of MOSFETs which has a very low power consumption. This is particularly important in battery- powered applications and in cases where heat dissipation is a problem.

collector one region of a bipolar transistor. In an npn transistor the electrons flow into this region when the device is switched ON.

conduction band the range of energies in a semiconductor in which an electron is able to move relatively easily through a crystal, and therefore contribute towards the process of electrical conduction. An electron is this band is called a conduction electron.

Cooper pairs a term describing the paired electrons which form in a superconductor.

covalent bond a bond which forms as the result of electrons being shared between two atoms.

critical temperature usually abbreviated by T_c—the temperature below which a material becomes a superconductor. The recently discovered materials with critical temperatures in excess of about 30 Kelvin are referred to as high T_c superconductors.

critical thickness   the maximum thickness of a layer of strained material which can be grown before the crystal structure is disrupted.

crystal   the regular arrangement of atoms which is present in virtually all solids.

current density   a measure of the amount of current flowing through a wire divided by the cross-sectional area of the wire.

CVD   or chemical vapour deposition —one method of forming atomically layered structures.

density of states   a measure of the number of electrons allowed over a given small energy range. The density of states is very low at the extremes of the valence and conduction bands and increases towards the middle of these bands.

depletion layer   the region around a p-n junction which is depleted of carriers. This forms because the conduction electrons and holes in this region have recombined as a result of the diffusion of carriers across the interface.

diffusion   a term applied to many situations in which the random movement of particles tends to produce a uniform distribution of the particles. In this book it applies mainly to the movement of conduction electrons and holes from regions of high concentration to areas with lower concentrations. At a p-n junction this gives rise to a diffusion current.

direct gap semiconductor   a material in which the recombination of conduction electrons and holes is an efficient process.

dislocation   a fault line which runs through a crystal. The existence of dislocations in a semiconductor is particularly detrimental to the behaviour of a device because conduction electrons tend to be trapped by the atoms which have a deficiency of electrons (see Figure 8.4).

donor   an impurity which has more valence electrons than the atom that it replaces. This generally gives rise to a conduction electron.

doping   the process of introducing donors or acceptors into a semiconductor in order to change the conductivity of the material.

double heterostructure laser   a laser in which the active region is enclosed between the two heterojunctions. The distinction from a quantum well laser is that the wavelike nature of the electrons and holes does not play a significant role.

drain   the region of a MOSFET through which the carriers flow out of the device.

effective mass   an electron in a crystal appears to be accelerated more rapidly by an electric field than we would expect. We explain this by saying that the electron has an effective mass. It is important to remember that this is a property of the crystal, not of the electron.

electric field strength   this is determined by the change in voltage over a given distance. Since we can picture the effect of a voltage as tilting the energy bands in a crystal, the electric field strength is a measure of the degree of tilt produced in a given structure.

electromigration   the physical destruction of a wire caused by the movement of ions from their position in the crystal. The effect is typically caused when a narrow wire is subject to a large current density.

electron–hole pair   the excitation of a valence electron into the conduction band leaves behind a hole in the valence band. Thus the two carriers are created simultaneously.

emitter   one region of a bipolar transistor. In an npn transistor the electrons flow out of this region into the base.

exciton   a system composed of a conduction electron and a hole which are attracted together by their opposing electrical charges and therefore move through the crystal as a single entity.

exclusion principle   a fundamental rule in quantum theory. For our purposes it states that no more than two electrons are allowed to occupy a given quantum state at any one time.

extrinsic semiconductor   a semiconductor in which dopant atoms are added to control the conductivity of the material.

Fermi energy   a quantity which is characteristic of the energy of the electrons in a material. At absolute zero the Fermi energy corresponds to the highest energy electrons, i.e. all the states with lower energy are occupied by electrons and all the ones with higher energy are vacant. At higher temperatures the Fermi energy corresponds to the average maximum energy of the electrons, i.e. there are as many electrons above the Fermi energy as there are vacant states below it.

FET   the acronym for field effect transistor, and the generic name for a number of devices using similar principles of operation.

gate   the central region of an FET.

Hall effect   the appearance of a voltage between the sides of a sample when simultaneously subjected to electric and magnetic fields applied at right-angles to one another.

HEMT   the acronym for high electron mobility transistor. These are also called MODFETs.

heterostructure   a device consisting of two (or more) different types of material. A heterojunction is the interface between these two materials.

hole   a particle which corresponds to the absence of an electron in the valence band of a semiconductor. The particle behaves as though it carries a positive charge. In most semiconductors the holes at the top of the valence band can have one of two different values of effective mass, and are referred to as light holes and heavy holes. Since their mobility depends on their effective mass, it is often necessary to distinguish between the two.

impurity scattering   the presence of impurities in a sample deflects the electrons and therefore gives rise to resistance. This is particularly important when the impurities form ions.

indirect gap semiconductor   a material in which the recombination of conduction electrons and holes occurs by an inefficient process.

interference a phenomenon which is characteristic of waves. When the waves are in phase the interference is constructive and the amplitude of the resultant waves is increased. When the waves are out of phase the interference is destructive and the waves cancel each other out.

intrinsic semiconductor a pure semiconductor containing no impurities. The only carriers present are thermally created conduction electrons and holes.

ion an atom which has gained or lost one or more electrons, and therefore has a net electric charge. Ionic bonding occurs in materials where the electrons are transferred from one type of atom to another. The bond is a result of the attractive forces between the positive and negative ions.

ion implantation a method of introducing dopant atoms into a crystal by projecting them at high speed towards the crystal surface.

Josephson junction a device formed by placing a very thin layer of insulator between two superconductors. The insulator is said to form a weak superconductor because the flow of superconducting current across the junction is extremely sensitive to a magnetic field.

majority/minority carriers in a semiconductor the majority carrier is the one introduced by doping the material. The minority carriers are the other type of carrier (conduction electrons or holes) present because of thermal excitation.

MBE the acronym for molecular beam epitaxy, which is one way of producing high-quality atomically layered structures.

MESFET the name given to an FET structure fabricated from gallium arsenide. The principle of operation is slightly different from that of a MOSFET because there is no oxide layer above the gate (since gallium arsenide has no native oxide).

metallic bonding the bond which forms between a collection of positive ions as a result of the associated sea of electrons. Materials which bond in this way are called metals. Owing to the large number of essentially free electrons, these materials make good electrical conductors.

mobility a measure of the ease with which a carrier can move through a given crystal.

modulation doping confinement of the doping to just one portion of a crystal so that carriers in the other regions experience high levels of mobility. An FET structure which makes use of this is the MODFET or modulation doped field effect transistor.

Moore's law a prediction made in 1964 by Gordon Moore that the number of devices in an integrated circuit would continue to double every year.

MOSFET the standard type of FET, and for many years the most commonly used transistor in integrated circuits. The initials stand for metal oxide semiconductor field effect transistor.

nipi superlattice a superlattice consisting of regions which are alternately doped with donors and acceptors with intervening intrinsic regions. The sequence of layers is then n-type, intrinsic, p-type, intrinsic, giving rise to the name nipi.

n-type semiconductor a material doped with donor impurities. The majority carriers are therefore conduction electrons.

periodic table a system in which the chemical elements are classified according to certain properties. In particular, the elements in Group I have one valence electron, those in Group II have two valence electrons, etc.

phase a term describing the specific stage that a wave is at in its cycle. In general we are interested in determining the phase of one wave relative to another. If the peaks of the waves coincide we say that they are in phase, if the peak of one coincides with the trough of the other then they are out of phase.

photolithography the process which is most commonly used to transfer the layout of an integrated circuit on to a wafer.

Planck's constant the fundamental constant associated with quantum theory. The energy of a photon is equal to Planck's constant multiplied by the speed of light and divided by the wavelength of the light.

p-n junction a device formed from a single crystal containing regions which are doped p-type and n-type. Among other applications the device can be used as a diode or in a p-n junction laser.

p-type semiconductor a material doped with acceptor impurities. The majority carriers are therefore holes.

quantum dot an artificial structure in which the carriers exhibit wavelike properties along all three dimensions.

quantum Hall effect the anomalous results obtained from the Hall effect when the carriers are confined in one dimension.

quantum well an artificial structure in which the carriers are confined in one dimension. In other words the electrons exhibit wavelike properties in one dimension but behave as free electrons in the other two dimensions. A quantum well laser uses these properties to produce a semiconductor laser which is far more efficient than a p-n junction laser.

quantum wire an artificial structure in which carriers are confined in two dimensions. The carriers are free to travel along the axis of the wire, but exhibit wavelike properties in the other directions.

recombination the process by which a conduction electron is reunited with a hole. As a result both carriers are annihilated and there is a release of energy (usually in the form of a photon).

resonant tunnelling a process in which the probability of an electron tunnelling through a barrier increases dramatically for a specific energy. This principle is applied in the resonant tunnelling transistor.

SEED the acronym for self electo-optic device—an optical switch which uses internal feedback to produce a resonant condition at a certain critical intensity.

single electron tunnelling the discrete flow of electrons tunnelling through an insulator.

source   the region of an FET through which the carriers flow into the device.

spontaneous emission   a randomly occurring process in which a conduction electron recombines with a hole to produce a photon.

stimulated emission   a process in which the presence of one photon induces recombination of an electron and hole to produce another identical photon.

strain   the result of producing a heterojunction between two materials with different lattice constants.

superconductor   a state of matter, generally achieved at extremely low temperatures, in which (among other things) the material exhibits no resistance to the flow of a direct current.

superlattice   an artificial structure formed by constructing a large number of quantum wells side by side, separated by thin barrier layers. Interaction of electrons in neighbouring wells produces a series of allowed minibands.

valence electrons   those electrons in the outermost orbit of an atom. On forming a solid these electrons interact to produce the valence band. In a semiconductor or insulator this band is full.

VLSI   the acronym for very large-scale integration, which describes integrated circuits with between a hundred thousand and ten million components.

FURTHER READING

Below is a list of books and articles in popular scientific journals which provide further information on the topics covered in this book. Some include a small amount of mathematics, but nothing too challenging.

Chapter 1

André Guinier, *The Structure of Matter: from the Blue Sky to Liquid Crystals* (Arnold, 1980). A very readable account which describes how the atomic view of matter is essential to our understanding of the physical properties of gases, liquids and solids

Chapters 2, 3 & 4

There are a great many books on semiconductors and semiconductor devices. Most also discuss the mobility of electrons in solids (Chapter 6) and the properties of superconductors (Chapter 10). My own personal favourite is:

L. Solymar and D Walsh, *Lectures on the Electrical Properties of Materials* (Oxford University Press, 1993)

Two other alternatives at a similar level are:

André Guinier and Remi Jullien, *The Solid State: from Superconductors to Superalloys* (Oxford University Press, 1989)

H. M. Rosenberg, *The Solid State* (Oxford University Press, 1978)

Chapter 5

William G. Oldham, 'The fabrication of microelectronic circuits', *Scientific American*, September 1977, p.111

Marc D. Levenson, 'Wavefront engineering for photolithography', *Physics Today*, July 1993, vol.**46**, p.28

James D. Meindl, 'Chips for advanced computing', *Scientific American*, October 1987, p.54

Chapter 6

W. R. Frensley, 'Gallium arsenide transistors', *Scientific American*, August 1987, p.68

M. H. Brodsky, 'Progress in gallium arsenide semiconductors', *Scientific American*, February 1990, p.56

Hadis Morkoc and Paul M. Solomon, 'The HEMT: a superfast transistor', *IEEE Spectrum*, February 1984, p.28

Trudy E. Bell, 'The quest for ballistic action', *IEEE Spectrum*, February 1986, p.36

Chapter 7

M. B. Panish, 'Molecular beam epitaxy', *Science*, 23 May 1980, vol.**208**, p.916

H. L. Stormer and D. C. Tsui, 'The quantized Hall effect', *Science*, 17 June 1983, vol.**220**, p.1241

Betrand I. Halperin, 'The quantized Hall effect', *Scientific American*, April 1986, p. 40

J. P. Eisenstein and H. L. Stormer, 'The fractional quantum Hall effect', *Science*, 22 June 1990, vol.**248**, p.1510

Chapters 8 & 9

Gottfried H. Döhler, 'Solid-state superlattices', *Scientific American*, November 1983, p.118

Federico Capasso, 'Band-gap engineering: from physics and materials to new semiconductor devices', *Science*, 9 January 1987, vol.**235**, p.172

Mani Sundaram, Scott A. Chalmers, Peter F. Hopkins and Arthur C. Gossard, 'New quantum structures', *Science*, 29 November 1991, vol.**254**, p.1326

V. Narayanamurti, 'Artificially structured thin-film materials and interfaces', *Science*, 27 February 1987, vol.**235**, p.1023.

Leroy L. Chang and Leo Esaki, 'Semiconductor quantum heterostructures', *Physics Today*, October 1992, vol.**45**, p.36

Mark A. Reed, 'Quantum dots', *Scientific American*, January 1993, p.98

Marc A. Kastner, 'Artificial atoms', *Physics Today*, January 1993, vol.**46**, p.24

Chapter 10

Randy Simon and Andrew Smith, *Superconductors: conquering technology's new frontier* (Plenum Press, 1988)—this book not only describes the physics of superconductors in layman's terms, but also covers an enormous range of applications of these materials

Juri Matisoo, 'The superconducting computer', *Scientific American*, May 1980, p.38

Karen Fitzgerald, 'Superconductivity: fact vs. fancy', *IEEE Spectrum*, May 1988, p.30

Physics Today, June 1991, vol. **44**, Special edition on high-temperature superconductors, including some applications to electronic devices

Kees Harmans, 'Next electron, please . . .', *Physics World*, March 1992, p.50

Konstantin K. Likharev and Tord Claeson, 'Single electronics', *Scientific American*, June 1992, p.50

Chapter 11

Dror G. Feitelson, *Optical Computing: a Survey for Computer Scientists* (MIT Press, 1988). Don't be misled by the title: this book is also of great interest to physicists and electrical engineers. It covers a wide range of different approaches to optical computing.

Eitan Abraham, Colin T. Seaton and S. Desmond Smith, 'The optical computer', *Scientific American*, February 1983, p.63.

S. D. Smith, 'Lasers, nonlinear optics and optical computers', *Nature*, 25 July 1985, vol.**316**, p.319

Trudy E. Bell, 'Optical computing: a field in flux', *IEEE Spectrum*, August 1986, p. 34

Chapter 12

Robert W. Keyes, 'The future of solid-state electronics', *Physics Today*, August 1992, vol.**45**, p.42.

In addition, a special edition of *Physics Today* (vol.**46**, June 1993) covers many of the subject areas considered in the later chapters. These are also treated in a more mathematical way in the following books:

M. Jaros, *Physics and Applications of Semiconductor Microstructures* (Oxford University Press, 1989)

Claude Weisbuch and Borge Vinter, *Quantum Semiconductor Structures: Fundamentals and Applications* (Academic Press, 1991)

Index